T0237200

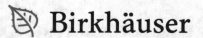

Mathematics and the Built Environment

Volume 5

Series Editors

Michael Ostwald (iD), Built Environment, University of New South Wales, Sydney, NSW, Australia

Kim Williams, Kim Williams Books, Torino, Italy

Edited by Kim Williams and Michael Ostwald.

Throughout history a rich and complex relationship has developed between mathematics and the various disciplines that design, analyse, construct and maintain the built environment.

This book series seeks to highlight the multifaceted connections between the disciplines of mathematics and architecture, through the publication of monographs that develop classical and contemporary mathematical themes – geometry, algebra, calculation, modelling. These themes may be expanded in architecture of any era, culture or style, from Ancient Greek and Rome, through the Renaissance and Baroque, to Modernism and computational and parametric design. Selected aspects of urban design, architectural conservation and engineering design that are relevant for architecture may also be included in the series.

Regardless of whether books in this series are focused on specific architectural or mathematical themes, the intention is to support detailed and rigorous explorations of the history, theory and design of the mathematical aspects of built environment.

More information about this series at http://www.springer.com/series/15181

Alberto Lastra

Parametric Geometry of Curves and Surfaces

Architectural Form-Finding

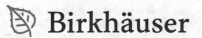 Birkhäuser

Alberto Lastra (iD)
Departamento de Física y Matemáticas
Universidad de Alcalá
Alcalá de Henares, Spain

ISSN 2512-157X   ISSN 2512-1561   (electronic)
Mathematics and the Built Environment
ISBN 978-3-030-81319-2   ISBN 978-3-030-81317-8   (eBook)
https://doi.org/10.1007/978-3-030-81317-8

Mathematics Subject Classification: 53A04, 53A05, 00A67

© The Editor(s) (if applicable) and The Author(s), under exclusive license to Springer Nature Switzerland AG 2021

This work is subject to copyright. All rights are solely and exclusively licensed by the Publisher, whether the whole or part of the material is concerned, specifically the rights of translation, reprinting, reuse of illustrations, recitation, broadcasting, reproduction on microfilms or in any other physical way, and transmission or information storage and retrieval, electronic adaptation, computer software, or by similar or dissimilar methodology now known or hereafter developed.

The use of general descriptive names, registered names, trademarks, service marks, etc. in this publication does not imply, even in the absence of a specific statement, that such names are exempt from the relevant protective laws and regulations and therefore free for general use.

The publisher, the authors, and the editors are safe to assume that the advice and information in this book are believed to be true and accurate at the date of publication. Neither the publisher nor the authors or the editors give a warranty, expressed or implied, with respect to the material contained herein or for any errors or omissions that may have been made. The publisher remains neutral with regard to jurisdictional claims in published maps and institutional affiliations.

This book is published under the imprint Birkhäuser, www.birkhauser-science.com, by the registered company Springer Nature Switzerland AG.
The registered company address is: Gewerbestrasse 11, 6330 Cham, Switzerland

A mis abuelos

Preface

The parametric aspects of curves and surfaces have been studied from the point of view of differential geometry through history. Indeed, many different studies have been developed since the nineteenth century on this discipline, which can be found in detail in texts such as do Carmo (1976), Tapp (2016), Umehara et al. (2017). Apart from the theoretical relation of a curve (or surface) to any of its parametrizations, one can go a step further and describe it from a practical point of view. The geometric scheme of a curve or a surface has provided inspiration for numerous works of art and architecture. These creations not only respond to physical needs such as certain acoustic properties, lighting, etc., but also to a human desire to create structures with simple geometric shapes. This books describes the classical theory of parametric tools in the geometry of curves and surfaces with an emphasis on applications to architecture.

This book is based on a decade of teaching a class on geometry in architectural studies at Universidad de Alcalá (Spain). This class, "Drawing Workshop II", combined architectural design and mathematics. Nevertheless, it can also be used as a text on differential geometry for mathematics students, or a basic reference on mathematics for architects and designers (especially those working with CAD). I hope that the latter will find this text useful and interesting, shedding light on the theoretical aspects of their work, as well as on the applicability to architecture. The techniques used in the examples provided in the text serve as the mathematical realization of many geometric tools used in CAD programs such as the construction of an helix, extrusions, revolution or ruled surfaces, projections, and many others. I also provide algorithms related to some of the geometric objects and show how different actions on the parametrizations change the nature of the geometric object itself. These mathematical tools are important to understand the structure of a geometric object and to know how to modify it consciously.

The structure of the book is as follows.

The first chapter is devoted to the study of parametrization of plane curves, with special focus on conics. In this chapter, the implicitation and approximation of curves is also illustrated with geometric examples.

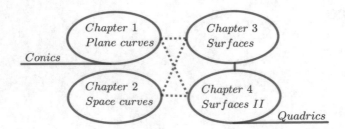

Fig. 1 Structure of the book

The second chapter describes a parallel theory on space curves, and the appearance of such curves in architecture. Geometric transformations are performed on a space curve, making explicit the mathematics behind usual actions in curve design.

The third and fourth chapters consider, respectively, general surfaces and other particular classes of surfaces which are of widespread use. The examples range from classic surfaces in architecture to the parametrization of such families or the construction of other which remain of particular interest regarding their properties. More precisely, we focus on some curves lying on surfaces and on the intersection of curves. Surfaces such as quadrics, ruled surfaces, surfaces of revolution, minimal and developable surfaces are also studied and applied to architectural elements.

The structure of the book is illustrated in Fig. 1.

The mathematical prerequisites for this book are first courses in topology, linear algebra and calculus (both single and multi- variable), as amply covered in the books Salas et al. (2003), Marsden and Tromba (2012), and Lang (1986); Strang (1993). For completeness, we have included two appendices covering knowledge that will be useful for understanding the material.

The figures of geometric objects have been created with Geogebra software.

I want to express my gratitude to everyone who was involved in this class, specially to Prof. Manuel de Miguel, who introduced me to the world of architecture. I also want to express my gratitude to Remi Lodh, who has guided me on its publication with high professionalism and also to Kim Williams for her enthusiasm, professionalism and effort in the revision of the manuscript, and also giving relevant and interesting details.

Suggested Further Reading

The following sources are suggested to interested readers seeking additional material (AAG 2008, 2010, 2013, 2014, 2016, 2018; Bridges 2003, 2004, 2008, 2011, 2012, 2014, 2016, 2018).

Alcalá de Henares, Spain   Alberto Lastra
2021

Contents

About the Author

Alberto Lastra has a Ph.D. in Mathematics by the University of Valladolid. He is associate professor at the University of Alcalá (Spain). He has been teaching mathematics in the degree of Architecture and Fundamentals in Architecture and Urbanism at the University of Alcalá since 2011, in subjects under the point of view of innovation and interdisciplinary thinking in Architecture. His research interests do not only go in the previous direction, but also in the study of asymptotic analysis of functional equations in the complex domain and related topics, symbolic computation, or orthogonal polynomials. He is a member of the research groups ECSING-AFA of the University of Valladolid and ASYNACS (CT-CE2019/683) of the University of Alcalá. He has also been a visitor at foreign research centers during the last decade, such as the University of Lille (France), the University of Warsaw (Poland), the University of La Rochelle (France), Universidade Federal de Minas Gerais (Brazil), among others.

List of Symbols

A^T	Transpose matrix of the matrix A
$\mathcal{C}^\infty(U)$	Set of scalar or vector functions which are differentiable for every degree of differentiation in the open set U
$d(P, Q)$	Euclidean distance from a point $P \in \mathbb{R}^n$ to $Q \in \mathbb{R}^n$
$dX_{(u_0, v_0)}(v)$	differential of $X : U \subseteq \mathbb{R}^2 \to \mathbb{R}^3$ at $(u_0, v_0) \in U$, evaluated at $v \in \mathbb{R}^2$
$D(P, r)$	Disc centered at the point P and radius $r > 0$
$\frac{d}{dx}$	derivative with respect to the variable
$\frac{\partial}{\partial x}, \frac{\partial}{\partial y}, \frac{\partial}{\partial z}, \dots$	Partial derivative with respect to $x, y, z, \dots$
$\mathcal{M}_{m \times n}(\mathbb{K})$	Set of $m \times n$ matrices with coefficients in a field $\mathbb{K}$
$I(\omega_1, \omega_2)$	First fundamental form
$II(\omega)$	Second fundamental form
$\nabla f(P)$	Gradient of the function f, evaluated at the point P
$\perp$	Orthogonal
$\mathrm{rank}(A)$	Rank of a matrix A
$\sim$	Asymptotic equivalence
$\langle \cdot, \cdot \rangle$ or $\cdot$	Inner product in $\mathbb{R}^n$
$\times$	Cross product in $\mathbb{R}^3$
$[\cdot, \cdot, \cdot]$	Scalar triple product
$\|\cdot\|$	Euclidean norm in $\mathbb{R}^n$
$\mathbb{C}$	Set of complex numbers
$\mathbb{Q}$	Set of rational numbers
$\mathbb{R}$	Set of real numbers
$\mathbb{R}^\star$	Set of real numbers, except from the origin. $\mathbb{R} \setminus \{0\}$
$\mathbb{Z}$	Set of integer numbers
$\mathbb{N}$	$\mathbb{Z} \setminus \{0, -1, -2, \dots\}$
$\vec{PQ}$	vector from the point P to the point Q of a Euclidean space
$\vec{v}$	vector of an Euclidean space
$\mathrm{Im}(f)$	Range of a function, i.e. $\{f(x) : x \in X\}$, whenever $f : X \to Y$
$\mathrm{Ker}(f)$	Kernel of a function $f : X \to \mathbb{R}^n$, i.e. $\{x \in X : f(x) = 0\}$
$\vec{v} \| \vec{w}$	The vectors $\vec{v}$ and $\vec{w}$ are parallel

List of Figures

Photograph Credits

Figure 1.5 Source: By Photo: Andreas Praefcke—Self-photographed, Public Domain,
Link: https://commons.wikimedia.org/w/index.php?curid=8382419

Figure 1.8 (left) Source: Mattancherry koonan kurish, Kochi, Kerala, India. Koonan Kurish Palli, Flickr. com.
Link: https://flic.kr/p/V4vMDm

Figure 1.8 (right) Source: Photo by Johnson Liu on Unsplash.
Link: https://unsplash.com/photos/C3SEO9ORkMg

Figure 1.10 Source: Photo by Luca Bravo on Unsplash
Link: https://unsplash.com/photos/alS7ewQ41M8

Figure 1.11 (left) Source: Photo by Andrea Junqueira on Unsplash
Link: https://unsplash.com/photos/mNoMLlDDJbg

Figure 1.11 (right) Source: Photo by Pavel Nekoranec on Unsplash
Link: https://unsplash.com/photos/-qJlgvKXE1M

Figure 1.21 Source: Public domain,
Link: https://commons.wikimedia.org/w/index.php?curid=37361

Figure 1.22 Source: Photo by Michael Martinelli on Unsplash.
Link: https://unsplash.com/photos/jgESEijOorE

Figure 1.23 Source: Photo by ckturistando on Unsplash.
Link: https://unsplash.com/photos/RWWHa5TUF8w

Figure 1.24 Source: Oceanogràfic, Wikipedia.com. De Felipe Gabaldón, CC BY 2.0,
Link: https://commons.wikimedia.org/w/index.php?curid=12532971

Figure 1.34 Source: Photo by Stephanie LeBlanc on Unsplash
Link: https://unsplash.com/photos/FiknH_A0SLE

Figue 1.37 Source: Photo stored in pxhere.com
Link: https://pxhere.com/es/photo/556035

Figure 1.38 Source: Photo stored in sphere.com
Link: https://pxhere.com/es/photo/556036

Figure 2.20 (left) Source: Photo by Yusuf Dündar on Unsplash
Link: https://unsplash.com/photos/Sm2IjyvrzDk

Figure 2.20 (right) Source: Photo by Steven Jackson, Gaudi's columns at Park Guell.Attribution 2.0 Generic (CC BY 2.0) On flic.kr
Link: https://flic.kr/p/9zsUoE

Figure 2.23 Source: Photo stored in pxhere.com
Link: https://pxhere.com/es/photo/555498

Figure 3.14 Source: Photo by Darcey Beau on Unsplash
Link: https://unsplash.com/photos/q8D7WZc40eA

Figure 3.18 Source: Photo by Nico Villanueva on Unsplash
Link: https://unsplash.com/photos/V89ZSyrExxs

Figures 3.21 and 3.22 Source: By Flickr user Hui Lan from Beijing, China—national theatre at Flickr, CC BY 2.0
Link: https://commons.wikimedia.org/w/index.php?curid=3334661

Figure 3.25 Source: Photo stored in pxhere.com
Link: https://pxhere.com/es/photo/1355308

Figure 3.27 (left) Source: Photo by Robby McCullough on Unsplash
Link: https://unsplash.com/photos/i7UsLKFX-Ms

Figure 3.27 (right) Source: Photo by Robby McCullough on Unsplash
Link: https://unsplash.com/photos/DtzJFYnFPJ8

Figure 4.11 (right) Source: Photo by Jonathan Singer on Unsplash
Link: https://unsplash.com/photos/Jda9U-CMc8c Source: Photo stored in pxhere.com

Figure 4.13 Source: By Unknown author—Maria Antonietta Crippa: Gaudí, Taschen, Köln, 2007, ISBN 978-3-8228-2519-8, Public Domain,
Link: https://commons.wikimedia.org/w/index.php?curid=3560968

Figure 4.14 Source: Photograph of the church of San Juan de Ávila, in Alcalá de Henares, taken on the 13th of February, 2021, by the author.

Figure 4.16 Source: De Andrés Franchi Ugart..., CC BY-SA 3.0
Link: https://commons.wikimedia.org/w/index.php?curid=54386268

Figure 4.40 Source: Photo stored in pxhere.com
Link: https://pxhere.com/es/photo/1042708

Figure 4.41 Source: Photo by Juliana Lee on Unsplash
Link: https://unsplash.com/photos/xibAcLZDUTY

Figure 4.42 Source: By Original uploader was Colin.faulkingham at en.wikipedia—Transfered from en.wikipedia Transfer was stated to be made by User:jcarkeys., Public Domain,
Link: https://commons.wikimedia.org/w/index.php?curid=3338724

Figure 4.43 Source: Photo stored in pxhere.com
Link: https://pxhere.com/es/photo/707570

Figure 4.45 Source: Photo stored in pxhere.com
Link: https://pxhere.com/es/photo/916411

Figure A.4 Source: Photo by Joel Filipe on Unsplash
Link: https://unsplash.com/photos/RFDP7_80v5A

Figure A.7 Source: Photo by Mevlüt ?ahin on Unsplash
Link: https://unsplash.com/photos/FevvUdyk-oE

Figure 3.30 Source: Wikipedia. Church of Kópavogur (icelandic: Kópavogskirk ja) being built ca. 1960
Link: https://is.m.wikipedia.org/wiki/Mynd:Kopavogur_church_ca1960.jpg

Figure C.25 Source: By M. Kocandrlova—Public Domain
Link: https://commons.wikimedia.org/w/index.php?curid=4327175

Chapter 1
Parametrizations and Plane Curves

This first chapter is devoted to the study of plane curves from the point of view of differential geometry. As mentioned in the preface, this theory is quite classical and can be found in different undergraduate and graduate textbooks such as do Carmo (1976), and more recent books (Tapp 2016; Umehara et al. 2017). The theory is illustrated with examples in specific architectural elements, together with mathematical techniques applied on plane curves. Among them, we focus on implicitation, approximation and interpolation techniques.

Different approaches might be followed in order to reach an adequate and consistent definition of curve: on the one hand, the point of departure can be located at the definition of a parametrized curve, arriving at the concept of the arc of a curve in implicit form, whilst other authors prefer to approach the concept of curve via the level curves of certain surfaces and conclude with local parametrizations. We will follow this second route, relating both approaches by means of the implicit function theorem. Certain topological discussions are important when going from one to the other point of view. For our purpose, we will only deal with curves which are locally homeomorphic to an open segment, i.e., such that at each of its points the curve is "essentially" a bent open interval near that point.

Both the implicit and parametric forms turn out to be important when handling a curve. The use of a parametrization of a curve is more manageable when trying to construct a curve by giving different values to the parameter, and the implicit form allows us to detect directly whether a point belongs to a given curve or not. Another advantage of having the implicit representation of a curve is the numerical application of the false position method (see Burden and Faires 2000, p. 72). Also, continuity of the function determining the curve in implicit form makes it possible to obtain slightly perturbed curves which assemble a plane curve determined by $f(x, y) = 0$ and the curve associated to the function $f(x, y) + c$, by considering small enough $c \in \mathbb{R}$.

Moreover, some elements related to a curve such as the curvature, tangent and normal vectors, etc. provide a wider knowledge of the curve. In addition to this,

© The Author(s), under exclusive license to Springer Nature Switzerland AG 2021
A. Lastra, *Parametric Geometry of Curves and Surfaces*, Mathematics and the Built Environment 5, https://doi.org/10.1007/978-3-030-81317-8_1

the definition of a curve itself may incorporate different physical or/and aesthetic properties such as building acoustics, the configuration of loads, architectural lighting, etc.

We have decided to consider the study of space curves separately to emphasize the importance of such curves, in particular helices, in architecture. As a matter of fact, the two previous approaches can be rephrased after natural adjustments.

1.1 Plane Curves and Parametrizations

We first state the concept of a plane curve, mainly consisting in the so-called level curve of a regular surface. Intuitively, a plane curve represents the path followed by a moving point inside a plane.

A circle and a lemniscate (see Fig. 1.1) are examples of the intuitive concept of a plane curve.

The circle consists of all points in the affine plane, $(x, y) \in \mathbb{R}^2$, whose components satisfy the equation $x^2 + y^2 = 9$:

$$C_1 = \{(x, y) \in \mathbb{R}^2 : x^2 + y^2 - 9 = 0\}.$$

The lemniscate corresponds to the set

$$C_2 = \{(x, y) \in \mathbb{R}^2 : (x^2 + y^2)^2 - xy = 0\}. \tag{1.1}$$

We define a plane curve following this approach, by means of a level curve of a scalar function f in two real variables.

Definition 1.1.1 Let $\emptyset \neq U \subseteq \mathbb{R}^2$ be an open set and let $f : U \to \mathbb{R}$ be a C^∞ function in U, i.e., a function that is differentiable for every degree of differentiation in U. We write $f \in C^\infty(U)$.

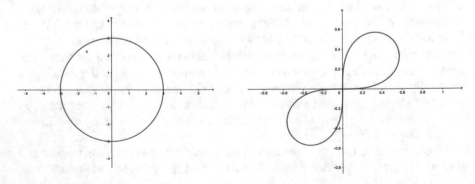

Fig. 1.1 Circle centered at $(0, 0)$ and radius $R = 3$ (left), lemniscate of Bernoulli (right)

In case the set

$$C = \{(x, y) \in U : f(x, y) = 0\}$$

is nonempty, we say that C is a **plane curve**.

A cardioid consists of the points $(x, y) \in \mathbb{R}^2$ such that

$$(x, y) = (2\cos(t) - \cos(2t), 2\sin(t) - \sin(2t)), \tag{1.2}$$

for some $t \in \mathbb{R}$. An epicycloid is a plane curve determined by the points $(x, y) \in \mathbb{R}^2$ satisfying

$$(x, y) = \left((r_1 + r_2)\cos(t) - r_2 \cos\left(\frac{r_1 + r_2}{r_2}t\right), (r_1 + r_2)\sin(t) - r_2 \sin\left(\frac{r_1 + r_2}{r_2}t\right) \right),$$

for some $t \in \mathbb{R}$, and where $r_1 \geq r_2$ are fixed positive real numbers.

Another approach to define a plane curve (or part of a plane curve) is by means of a parametrization of such curve, as we have already observed (see Fig. 1.2).

Definition 1.1.2 A **parametrization** is a pair (I, α), where $I \subseteq \mathbb{R}$ is an open interval and $\alpha : I \to \mathbb{R}^2$ is a C^∞ function on I, i.e. $\alpha \in C^\infty(I)$.

It is worth pointing out that the hypothesis of f belonging to the set $C^\infty(I)$ is too restrictive for the underlying theory. As a matter of fact, the previous definition can be stated on other more general subsets $I \subset \mathbb{R}$. We have decided to maintain such restrictions for the sake of simplicity in our reasoning, and place an emphasis on the applications in architecture rather than the accuracy on the hypotheses made on the results.

We will mainly work with regular parametrizations.

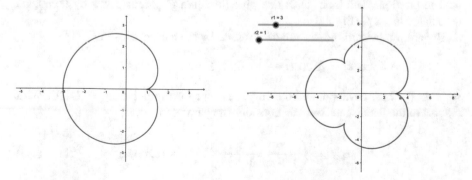

Fig. 1.2 Cardioid (left) and epicycloid (right)

Definition 1.1.3 A parametrization (I, α) is a **regular parametrization** if it satisfies the following conditions:

- $\alpha'(t) \neq (0, 0)$ for all $t \in I$;
- $\alpha : I \to \mathbb{R}^2$ is a one-to-one function.

The first statement in Definition 1.1.3 allows us to define the tangent line at each point of the curve associated to the regular parametrization.

In the example of the cardioid, we have that

$$\alpha(t) = (2\cos(t) - \cos(2t), 2\sin(t) - \sin(2t)),$$

$t \in (-\pi, \pi)$ covers the whole curve except from the point $P_1 = (-3, 0)$. Observe that

$$\alpha'(t) = (-2\sin(t) + 2\sin(2t), 2\cos(t) - 2\cos(2t)),$$

therefore $\alpha'(0) = (0, 0)$.

The property of injectivity associated to a regular parametrization is needed in order not to observe self-intersection points in the curve, as happened in the example of the lemniscate.

In order to fulfill the two conditions in Definition 1.1.3, the use of **local parametrizations** of curves is useful. So far, we have defined plane curves by means of the level curves of some function in two variables. Given a parametrization, it is natural to think that the image of a parametrization determines a curve (or part of it).

Definition 1.1.4 Given a plane curve C, we say that the parametrization (I, α) is a **parametrization of the curve** C if $C = \alpha(I)$.

The image set of a parametrization (I, α), $\alpha(I)$, is known as an **arc**. An arc is said to be **regular** if it is described by a regular parametrization.

The previous definition states that given a curve C parametrized by (I, α), its associated arc is C (Fig. 1.3).

Regarding the cardioid, the arc determined by the parametrization

$$\alpha(t) = (2\cos(t) - \cos(2t), 2\sin(t) - \sin(2t)),$$

for $t \in (0, 2\pi)$, covers the whole curve, except for the point $P_2 = (1, 0)$. With respect to the lemniscate, one can consider the parametrizations

$$\alpha_1(t) = \left(\frac{t}{1 + t^4}, \frac{t^3}{1 + t^4} \right), \quad t \in (0, \infty), \tag{1.3}$$

$$\alpha_2(t) = \left(\frac{t}{1 + t^4}, \frac{t^3}{1 + t^4} \right), \quad t \in (-\infty, 0), \tag{1.4}$$

which together draw the whole curve, except for the origin.

Fig. 1.3 Regular curve

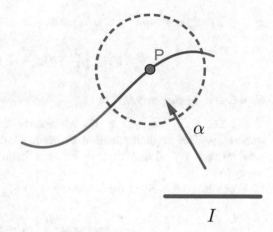

The previous examples motivate the following definition of a regular curve.

Definition 1.1.5 A set $\emptyset \neq C \subseteq \mathbb{R}^2$ is a **regular (plane) curve** if for every $(x_0, y_0) \in C$ there exists a disc $D((x_0, y_0), r) \subseteq \mathbb{R}^2$, such that $D((x_0, y_0), r) \cap C$ is a regular arc.

The following result is a direct application of the implicit function theorem.

Theorem 1.1.6 *Let $U \subseteq \mathbb{R}^2$ be a nonempty open set of $\mathbb{R}^2$. Let $f : U \to \mathbb{R}$ with $f \in C^\infty(U)$ and $(x_0, y_0) \in U$ such that*

$$\nabla f(x_0, y_0) = \left(\frac{\partial f}{\partial x}(x_0, y_0), \frac{\partial f}{\partial y}(x_0, y_0) \right) \neq (0, 0).$$

We consider the set

$$C = \{(x, y) \in U : f(x, y) - f(x_0, y_0) = 0\} \neq \emptyset.$$

Then, there exist $r > 0$, an open interval $I \subseteq \mathbb{R}$ and a function $\alpha : I \to \mathbb{R}^2$, $\alpha \in C^\infty(I)$ such that

- *$\alpha'(t) \neq (0, 0)$ for all $t \in I$,*
- *$\alpha : I \to \mathbb{R}^2$ is a one-to-one function, and $\alpha(I) = D((x_0, y_0), r) \cap C$.*

In terms of the concepts introduced above, Theorem 1.1.6 can be stated as follows:

Theorem 1.1.7 *Let $U \subseteq \mathbb{R}^2$ be a nonempty open set of $\mathbb{R}^2$. Let $f : U \to \mathbb{R}$ with $f \in C^\infty(U)$. Consider the set*

$$C = \{(x, y) \in U : f(x, y) = 0\}.$$

If $C \neq \emptyset$ and for every $P \in C$ it holds that

$$\nabla f(P) = \left(\frac{\partial f}{\partial x}(P), \frac{\partial f}{\partial y}(P) \right) \neq (0, 0),$$

then C is a regular curve.

Proof Let $P = (x_0, y_0) \in C$. We assume that $\frac{\partial f}{\partial x}(P) \neq 0$, without loss of generality. The implicit function theorem guarantees the existence of an open interval $x_0 \in I \subseteq \mathbb{R}$ and $\alpha : I \to \mathbb{R}^2$ such that $\alpha(x_0) = y_0$ and $f(t, \alpha(t)) = 0$ for every $t \in I$.

From the continuity of the function $t \in I \mapsto \frac{\partial f}{\partial x}(t, \alpha(t))$ and the fact that

$$\frac{\partial f}{\partial x}(x_0, y_0) = \frac{\partial f}{\partial x}(x_0, \alpha(x_0)) \neq 0,$$

there exists $x_0 \in I_1 \subseteq I$ such that $\frac{\partial f}{\partial x}(t, \alpha(t)) \neq 0$ for all $t \in I_1$.

- From the construction of α we obtain that $(t, \alpha(t)) \in C$ for all $t \in I_1$.
- $\alpha'(t) \neq 0$ for all $t \in I_1$. Otherwise, assume the existence of some $t_0 \in I_1$ with $\alpha'(t_0) = 0$. Taking derivatives in $f(t, \alpha(t)) = 0$ we obtain that

$$\frac{\partial f}{\partial x}(t, \alpha(t)) + \frac{\partial f}{\partial y}(t, \alpha(t))\alpha'(t) \equiv 0, \quad t \in I_1.$$

This yields that $\frac{\partial f}{\partial x}(t_0, \alpha(t_0)) = 0$, which contradicts the choice of I_1. Therefore, $\alpha'(t) \neq 0$ for all $t \in I_1$.

- $\alpha : I_1 \to \mathbb{R}^2$ is a one-to-one function. If there exist $t_1, t_2 \in I_1$ such that $\alpha(t_1) = \alpha(t_2)$, then, Rolle theorem guarantees the existence of $t_3 \in I_1$ with $\alpha'(t_3) = 0$ which contradicts the previous statement.

We observe from the proof that an adequate parametrization can be considered for each point in the curve, the result being of a local nature. Moreover, the existence of a local parametrization is guaranteed by the implicit function theorem, depending on the component of the gradient which does not vanish.

A reciprocal local result is also valid, describing the reciprocal relationship between local regular parametrizations and level curves of scalar functions in two variables.

Theorem 1.1.8 *Let C be a regular curve. For every $(x_0, y_0) \in C$ there exist $r > 0$ and a scalar function $f : D((x_0, y_0), r) \to \mathbb{R}, f \in C^\infty(D((x_0, y_0), r))$, such that*

$$C \cap D((x_0, y_0), r) = \{(x, y) \in D((x_0, y_0), r) : f(x, y) = 0\},$$

and

$$\nabla f(Q) = \left(\frac{\partial f}{\partial x}(Q), \frac{\partial f}{\partial y}(Q)\right) \neq (0, 0),$$

for all $Q \in D((x_0, y_0), r)$.

Proof Let $(x_0, y_0) \in C$. We consider the disc $D((x_0, y_0), r_1) \subseteq \mathbb{R}^2$ such that $D((x_0, y_0), r_1) \cap C$ is an arc of regular curve. This entails the existence of a regular parametrization (I, α) with $\alpha(I) = D((x_0, y_0), r_1) \cap C$. Let us write $\alpha(t) = (\alpha_1(t), \alpha_2(t))$. Let $t_0 \in I$ with $\alpha(t_0) = (x_0, y_0)$ and assume, without loss of generality, that $\alpha_1'(t_0) \neq 0$.

From the continuity of α_1' in I, we can guarantee the existence of an open interval $\emptyset \neq I_1 \subseteq I$ in which $\alpha_1'(t) \neq 0$ for all $t \in I_1$. This entails that the function α_1 is invertible in I_1, with $\alpha_1(I_1) = I_2$ for some open interval I_2. It is not difficult to verify, reducing I_1 if necessary, the existence of $0 < r \leq r_1$ such that $(x_0, y_0) \in \alpha(I_1) = D((x_0, y_0), r) \cap C$. The pair (I_1, α) is a regular parametrization which parametrizes an arc contained in C.

Let $f : D((x_0, y_0), r) \to \mathbb{R}$ be defined by

$$f(x, y) = y - \alpha_2(\alpha_1^{-1}(x)).$$

Observe that f is well defined for all $(x, y) \in D((x_0, y_0), r_1)$ due to injectivity of α_1. Given $(x, y) \in D((x_0, y_0), r) \cap C$, we have $x = \alpha_1(t)$ and $y = \alpha_2(t)$, for some $t \in I_1$. Then, $t = \alpha_1^{-1}(x)$ and $y = \alpha_2(\alpha_1^{-1}(x))$. The function f is infinitely differentiable in $D((x_0, y_0), r)$. In addition to this, $\frac{\partial}{\partial y} f(x, y) = 1 \neq 0$.

The previous results give rise to the concept of regular curve, when the point of departure is an implicit expression.

Definition 1.1.9 Let $U \subseteq \mathbb{R}^2$ be a nonempty open set of $\mathbb{R}^2$. Let $f : U \to \mathbb{R}$ with $f \in C^\infty(U)$. We say that the set

$$C = \{(x, y) \in U : f(x, y) = 0\}$$

is a **regular curve** if $C \neq \emptyset$ and for all $P \in C$

$$\nabla f(P) = \left(\frac{\partial f}{\partial x}(P), \frac{\partial f}{\partial y}(P)\right) \neq (0, 0).$$

The existence of a disc for each point $(x_0, y_0) \in C$ such that the intersection of that disc and the curve coincides with the image of a regular parametrization is essential in the hypotheses of the definition of a regular curve given in terms of regular parametrizations (see Definition 1.1.5). The usual topology on $\mathbb{R}^2$ determines counterexamples in this direction.

Example 1.1.10 Let us consider the arc determined by the following parametrization. That arc is contained in the lemniscate described above in this chapter:

$$\alpha(t) = \left(\frac{t}{1+t^4}, \frac{t^3}{1+t^4}\right), \quad t \in (-1, \infty).$$

α turns out to be a $C^\infty(-1, \infty)$ function. The pair $((-1, \infty), \alpha)$ is a regular parametrization:

$$\alpha'(t) = \left(\frac{1-3t^4}{t^8 + 2t^4 + 1}, \frac{t^2(3-t^4)}{t^8 + 2t^4 + 1}\right) \neq (0,0), \quad t \in (-1, \infty),$$

- It is direct to verify that $\alpha'(t) \neq (0,0)$ for $t \in (-1, \infty)$.
- α is a one-to-one function.

However, for every disc centered at $\alpha(0) = (0,0)$, the set obtained by intersection of the curve and the disc can not be parametrized by any regular parametrization. That subset is not homeomorphic to a segment, i.e., it can not be transformed by continuous transformations into a segment (this set has two connected components whereas a segment has only one (Fig. 1.4)).

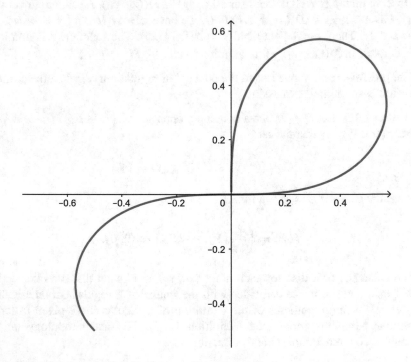

Fig. 1.4 Counterexample of regular curve in Example 1.1.10

1.2 Some Classic Curves in Architecture

In this section, we show how classic plane curves are used to describe and inspire architectural elements. This is not accidental, and is more likely due to some necessity in the structure, for aesthetic reasons, etc. In the book Hanh (2012), several particular studies are made on plane curves applied in architectural design.

Cycloid, and the Kimbell Art Museum
The Kimbell Art Museum (Forth Worth, Texas, 1972) by Louis Khan is a building of known geometric complexity. Its roof consists in the concatenation of several vaults built up from a plane curve and parallel lines passing through that curve (see Fig. 1.5).

Each of the vaults is based on the plane curve known as the **cycloid**. The cycloid is a plane curve defined by a physical phenomena. Let a circle roll on a line. The trail left by any fixed point in the circle after this movement draws a cycloid (see Fig. 1.5).

The equations defining a parametrization of the cycloid can be derived from the physical definition, making use of elementary trigonometry and fundamentals of physics (Fig. 1.6).

We write $r > 0$ for the radius of the rolling circle. Assume the initial position of the rolling circle is given by the equation $x^2 + (y - r)^2 = r^2$, and the distinguished

Fig. 1.5 Kimbell Art Museum

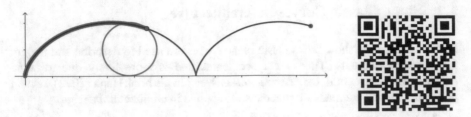

Fig. 1.6 Rolling circle producing a cycloid. QR Code 1

point is the origin of coordinates $P = (0, 0)$. Given any positive time $s > 0$, the rolling circle has been transformed into the circle of equations $(x - s)^2 + (y - r)^2 = r^2$. This entails that the point P moves clockwise inside the circle s units. The length of an arc in a circle is given by $r\alpha$, so $\alpha = s/r$. In conclusion, the initial angle of the distinguished point in the circle is $-\pi/2$, so the angle of P at the time s should be $-\pi/2 - s$. This yields that the possition of P at any positive time s is given by

$$\left(s + r\cos(-\frac{s}{r} - \frac{\pi}{2}), r(\sin(-\frac{s}{r} - \frac{\pi}{2}) + 1)\right).$$

From the trigonometric properties of the sum of an angle, and by the change of scale in the parameter $s = rt$, we derive the following parametrization of the cycloid.

$$\begin{cases} x(t) = r(t - \sin(t)) \\ y(t) = r(1 - \cos(t)) \end{cases} \quad t \in \mathbb{R}. \tag{1.5}$$

An implicit equation determining the cycloid is given by

$$C = \{(x, y) \in \mathbb{R}^2 : x - r\arccos(1 - \frac{y}{r}) + \sqrt{2yr - y^2} = 0\}.$$

Another physical property which can be derived from the geometry of this curve is the condition that the amount of time that it takes any object to fall from any height in a cycloid is equal to the time it would take it to reach the minimum point on the cycloid.

Example 1.2.1 The curvature of a cycloid, parametrized in in (1.5), for every value of the parameter t is given as follows. Let $(\mathbb{R}, \alpha)$ with $\alpha(t) = (r(t - \sin(t)), r(1 - \cos(t)))$ be the parametrization of the cycloid. In view of Proposition 1.3.20, we get that for all $t \notin \{2\pi k : k \in \mathbb{Z}\}$

$$\kappa_\alpha(t) = \frac{\begin{vmatrix} r(1 - \cos(t)) & r\sin(t) \\ r\sin(t) & r\cos(t) \end{vmatrix}}{(2r^2(1 - \cos(t)))^{3/2}} = \frac{-\sqrt{2}}{4r\sqrt{1 - \cos(t)}}.$$

Catenary Arch

We intuitively define a **catenary** to be the form described by a string when it is allowed to hang loosely from its extremes. by its extremes. The physical formulas described by the forces in this situation lead to a parametrization of the curve given by

$$
\begin{cases}
x(t) = t \\
y(t) = \frac{a}{2}\left(e^{\frac{t}{a}} + e^{-\frac{t}{a}}\right) & t \in \mathbb{R}
\end{cases}
, \tag{1.6}
$$

where a is a positive parameter describing the spacing between the fastening points.

The QR code in Fig. 1.7 draws different catenaries by changing the value of the parameter $a > 0$.

The main physical property of a catenary is that all the horizontal stresses are compensated for, and the figure remains stationary. The inverted catenary arch has been used in the Keleti railway station in Budapest designed by Gyula Rochlitz and János Feketeházy, and several works by Antoni Gaudí in Spain include this kind of archs. The thrust properties of a catenary are endowed by catenary arches

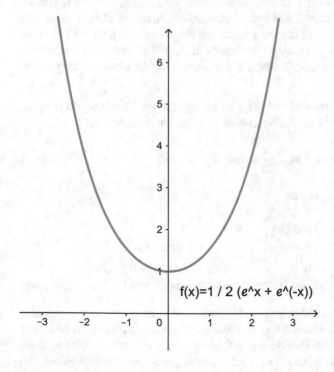

$$f(x) = 1 / 2 \; (e\char`^x + e\char`^(-x))$$

Fig. 1.7   Catenary, $a = 1$. QR Code 2

Fig. 1.8 Catenary archs

so that lateral thrusts are minimized because the arch supports itself. Therefore the geometric form of inverted catenary arch allows us to minimize the compressive stresses of the arch. Figure 1.8, left, illustrates St. George Orthodox Koonan Kurish Old Syrian Church in Mattancherry, India, while Fig. 1.8, right, shows the Gateway Arch in St. Louis, USA, by Eero Saarinen.

A catenary-like structure entitled "The Catène de Containers" designed by Vincent Ganivet which can be found in Le Havre is formed of two elements, each of which consists of a stacking of containers forming this shape. We also refer to Hanh (2012) for some details on catenary-like shapes in certain oeuvres in architecture, and to Hart and Heathfield (2018), where the authors show different catenary-based constructions.

Example 1.2.2 The curvature at any point of the catenary arch can be determined by a parametrization of this curve. We consider the parametrization in (1.6)

$$\alpha(t) = (t, \frac{a}{2}\cosh(\frac{t}{a})), \quad t \in \mathbb{R}$$

for some fixed $a > 0$. We get that

$$\kappa_\alpha(t) = \frac{\begin{vmatrix} 1 & \frac{1}{2}\sinh(\frac{t}{a}) \\ 0 & \frac{1}{2a}\cosh(\frac{t}{a}) \end{vmatrix}}{(1 + \sinh^2(t/a))^{3/2}} = \frac{\cosh(t/a)}{2a(1 + 1/4\sinh^2(t/a))^{3/2}}.$$

Lemniscate
Mathematical and physical advances made during the seventeenth and eighteenth centuries gave rise to architectural theories based on the refounding of the theory of proportions. In Gerbino (2014), the chapter by Filippo Camerota is devoted to these issues. There one can find the use of ellipses appearing in oblique deformations, and also catenary forms in domes or bridges, as well as cycloids.

A **lemniscate of Bernoulli** is a plane curve whose definition assembles that of the ordinary conics as places in the Euclidian plane. More precisely, it consists of the points in the plane such that the product of distances to two fixed points P_1, and P_2 (called foci), with $\text{dist}(P_1, P_2) = 2a > 0$, is given by a^2, for some fixed $a > 0$.

This definition allows us to determine an implicit equation defining the lemniscate. Let us assume that $P_1 = (-a, 0)$ and $P_2 = (a, 0)$. We search for all $(x, y) \in \mathbb{R}^2$ such that $\text{dist}((x, y), (-a, 0)) \cdot d((x, y), (a, 0)) = a^2$. The implicit equation is derived from there, arriving at

$$C = \{(x, y) \in \mathbb{R}^2 : (x^2 + y^2)^2 = 2a^2(x^2 - y^2)\}. \tag{1.7}$$

The expression

$$\begin{cases} x(t) = \frac{a\sqrt{2}\cos(t)}{\sin^2(t)+1} \\ y(t) = \frac{a\sqrt{2}\cos(t)\sin(t)}{\sin^2(t)+1} & t \in \mathbb{R} \end{cases}, \tag{1.8}$$

is a parametrization of such curve. Observe that the curve is drawn for every interval of length 2π.

In Echevarría et al. (2014), the authors study the structure of a chapel by Eduardo Torroja, located in Madrid, whose form is based on a lemniscate. Another application of the lemniscate can be found in Calió and Marchetti (2015), where the authors analyze the Sogn Benedetg Chapel, by Peter Zumthor, a building whose floor plan contains a lemniscate (Fig. 1.9).

Sinusoidal Curves

Sinusoidal curves may appear as sections with planes in certain surfaces defining elements in some buildings. Examples of this kind of curves appear in San Juan de Ávila Church and Cristo Obrero church by Eladio Dieste, located in Alcalá de Henares, Spain, and Atlántida, Uruguay, respectively (see Figs. 4.14 and 4.16, respectively). In both of them, the side walls form curves of equation $y - a\sin(bx) = 0$, for certain $a, b \in \mathbb{R}$, depending on the height.

Other examples of such curves are shown in Zentrum Paul Klee by Renzo Piano, located in Bern, Switzerland; the Mediopadana high-speed train station by Santiago Calatrava in the province of Reggio Emilia, Italy (see Fig. 1.10); the Centre Pompidou-Metz by Shigeru Ban in Metz, France; or the Altra Sede Regione Lombardia, Milano, Italy, by Pei Cobb & Partners with Caputo Partners.

Logarithmic Spirals

Logarithmic spirals usually appear due to a perspective phenomena related to buildings. In particular, a logarithmic spiral is due to the orthogonal projection of a

Fig. 1.9 Lemniscate

Fig. 1.10   The high-speed train station Reggio Emilia AV Mediopadana, Italy

Fig. 1.11   Spirals: Casa Batló, by Antoni Gaudí (left) and Staircase (right)

viewer's worm's eye view of a circular helix, that is, a helix which in the floor plan produces a circle (see Fig. 1.11, right). This is not a curve explicitly constructed to conform to the shape of the construction, but associated to it. Spirals also appear in the ceiling of Casa Batló by Antoni Gaudí in Barcelona (see Fig. 1.11, left).

In Sect. 2.5, we will give more detail on different actions on a helix, modifying the elements creating their characteristic form. That point will clarify several modifications made on spirals.

Fig. 1.12 Logarithmic spiral.
$a = 1, b = 0.3$

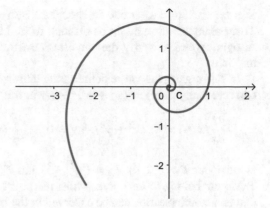

A logarithmic spiral is parametrized by

$$\begin{cases} x(t) = a \exp(bt) \cos(t) \\ y(t) = a \exp(bt) \sin(t) \qquad t \in \mathbb{R} \end{cases}, \tag{1.9}$$

for some $a, b \in \mathbb{R}$ (see Fig. 1.12). The modulus of the parameter $a \in \mathbb{R}$ determines the scale of the spiral, whereas its sign informs us about the direction in which the spiral opens to infinity (clockwise or counterclockwise). The parameter b distinguishes the separation rate of the spiral from the origin.

In the chapter by A. C. Huppert in Gerbino (2014), the author shows how the use of mathematical objects is important in the Italian Renaissance architecture, where forms such as circles and spirals often appear. Other spiralling patterns are included in Dunham (2003).

Other Plane Curves
Different manifestations of plane curves may appear in architectural elements due to physical or aesthetic demands. For example, the appearance of a troposkein in a building is due to the need to minimize bending stresses. As a matter of fact, its shape is that of a skipping rope, due to the interacting forces. We refer to Sharpe (2010) and the references therein for the use of such curves in existing buildings. Other curves resemble natural phenomena such as river meanders, as in the plan view of the Museum of the Human Body in Montpellier designed by Bjarke Ingels Group (BIG). See Langbein and Leopold (1970) for a mathematical insight and model of these curves.

1.3 Some Elements of Regular Plane Curves

Theorems 1.1.7 and 1.1.8 have linked the two previous manners of defining plane curves. As a matter of fact, Definition 1.1.5 is based on an existence argument. Locally, one has to verify that in a neighborhood of every point, the set corresponds to a regular arc, i.e., it can be described by means of a regular parametrization. That parametrization might be difficult to find in practice. In contrast to this, it is straightforward to verify the conditions on the gradient when a curve is defined implicitly.

In this regard, the curve defining the lemniscate in (1.1) is not a regular curve. Observe that $f(x, y) = (x^2 + y^2)^2 - xy$ in this example, and it holds that

$$\frac{\partial f}{\partial x}(x, y) = 4x(x^2 + y^2) - y = 0, \quad \frac{\partial f}{\partial y}(x, y) = 4y(x^2 + y^2) - x = 0$$

at the points $P_1 = (0, 0)$, $P_2 = (\frac{\sqrt{2}}{4}, \frac{\sqrt{2}}{4})$ and $P_3 = (\frac{-\sqrt{2}}{4}, \frac{-\sqrt{2}}{4})$. The points P_2 and P_3 do not belong to the curve, whilst the point $P_1 = (0, 0)$ is special in the sense that an autointersection can be observed in the graph of the lemniscate.

Regarding the regular parametrization of the cardioid in (1.2), one can verify that

$$x'(t) = -2\sin(t) + 2\sin(2t), \quad y'(t) = 2\cos(t) - 2\cos(2t),$$

which vanish at the values of the parametrer $t_k = 2\pi k$, for every $k \in \mathbb{Z}$. Observe that such a parametrization draws the whole lemniscate for each interval of length 2π. This means that the only controversial point is $P_1 = (0, 0)$. This specific parametrization is not regular at P_1. Observe that $(0, 0)$ has a particular shape in the cardioid (see Fig. 1.2, left). The point $P = (0, 0)$ is a **cusp** in the cardioid.

The theory behind cusps, and singular points in general, is vast and very interesting and is beyond the scope of this book. For a detailed study of such points, we refer to Section 7.5 in Gibson (2001).

An example of a cusp is that of the cardioid, whose parametrization at $t = 0$ satisfies the condition that both derivatives $x'(t)$ and $y'(t)$ vanish, and $(x''(t), y''(t)) \neq (0, 0)$. Generalizations on different orders of null derivatives provide other types of cusps. In the parametrization of the cardioid, one has

$$x''(t) = -2\cos(t) + 4\cos(2t), \quad y''(t) = -2\sin(t) + 4\sin(2t).$$

Observe that one of the previous second derivatives does not vanish at $t = 0$.

The last examples lead to the next definition.

Definition 1.3.1 Given a plane curve C, we say that $P \in C$ is a **singular point** of C if there does not exist $r > 0$ such that $D(P, r) \cap C$ is a regular arc, in other words, if C does not describe a regular curve in a vicinity of P.

A point P in a plane curve C is said to be **regular** if it is not a singular point.

Given a regular parametrization (I, α), it is interesting to consider for any $t_0 \in I$ the vector $\alpha'(t_0) \in \mathbb{R}^2$, which turns out to be different from $(0, 0) \in \mathbb{R}^2$.

Definition 1.3.2 Let (I, α) be a regular parametrization, and $t_0 \in I$, with $P = \alpha(t_0)$. The **velocity vector** (or tangent vector) associated to (I, α) at P is defined by $\alpha'(t_0) \in \mathbb{R}^2$.

The line at $P = \alpha(t_0)$ and direction vector given by its velocity vector $\alpha'(t_0)$ is known as the **tangent line** at P, associated to (I, α).

As a matter of fact, given a regular curve C, the tangent line at a point does not depend on the choice of the regular parametrization, so one can talk about the tangent line at P of the curve C.

Proposition 1.3.3 *The tangent line of a regular curve C at a point does not depend on the regular parametrization.*

Proof Let $P \in C$ and let (I_1, α_1) and (I_2, α_2) be two regular parametrizations of a regular arc of C with $P \in \alpha_1(I_1) \cap \alpha_2(I_2)$. Assume that $P = \alpha_1(t_1) = \alpha_2(t_2)$, for some $t_1 \in I_1$ and $t_2 \in I_2$. We only have to verify that the vectors $\alpha_1'(t_1)$ and $\alpha_2'(t_2)$ are proportional.

Lemma 1.3.4 *Let (I, α) be a regular parametrization. The tangent line at $P = \alpha(t_0)$ is the limit position of the secant lines of C crossing at P and $\alpha(t)$, when $\alpha(t)$ approaches $\alpha(t_0)$.*

Proof Taylor's theorem guarantees that

$$\alpha(t) = \alpha(t_0) + (t - t_0)\alpha'(t_0) + u(t),$$

with

$$\lim_{t \to t_0} \frac{u(t)}{t - t_0} = 0.$$

The line joining $\alpha(t)$ and $\alpha(t_0)$ has direction vector given by $\alpha(t) - \alpha(t_0)$ which is proportional to $\frac{\alpha(t) - \alpha(t_0)}{t - t_0}$. The previous line (Fig. 1.13) is given by the following equation:

$$(x, y) = \alpha(t_0) + t \frac{\alpha(t) - \alpha(t_0)}{t - t_0}.$$

Applying the limit $t \to t_0$ we obtain that the secant tends to the tangent line

$$(x, y) = \alpha(t_0) + t\alpha'(t_0).$$

The result follows directly from Lemma 1.3.4, which is only based on a geometric property which does not depend on the regular parametrization chosen.

$$(x(s'), y(s')) = (\alpha_1(t_0), \alpha_2(t_0)) + s' \left(\underbrace{\frac{\alpha_1(t) - \alpha_1(t_0)}{t - t_0}}_{\downarrow \atop \alpha_1'(t_0)}, \underbrace{\frac{\alpha_2(t) - \alpha_2(t_0)}{t - t_0}}_{\downarrow \atop \alpha_2'(t_0)} \right).$$

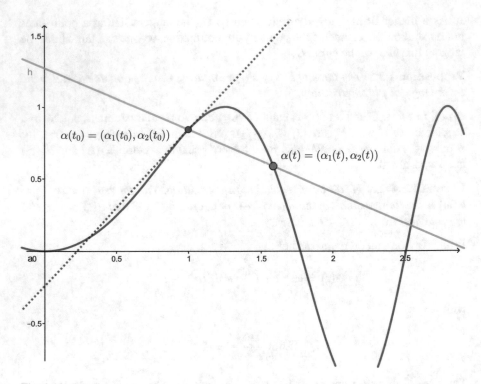

Fig. 1.13 Secant

The next definition now makes sense.

Definition 1.3.5 Let C be a regular curve and $P \in C$. For every regular parametrization (I, α) of an arc of curve in C with $P \in \alpha(I)$, the tangent line of C at P is parametrized by

$$(x(t), y(t)) = P + \alpha'(t_0)t, \quad t \in \mathbb{R},$$

with $P = \alpha(t_0)$.

Proposition 1.3.6 *Let* $C = \{(x, y) \in \mathbb{R}^2 : f(x, y) = 0\}$ *be a regular curve. For every* $P = (x_0, y_0) \in C$, *the tangent line to* C *at* P *is given by*

$$\frac{\partial f}{\partial x}(P)(x - x_0) + \frac{\partial f}{\partial y}(P)(y - y_0) = 0. \tag{1.10}$$

Proof Let $P \in C$ and let (I, α) be a regular parametrization of an arc of C with $P \in \alpha(I)$. Assume that $P = \alpha(t_0)$, for some $t_0 \in I$.

It holds that

$$f(\alpha(t)) = 0, \quad t \in I.$$

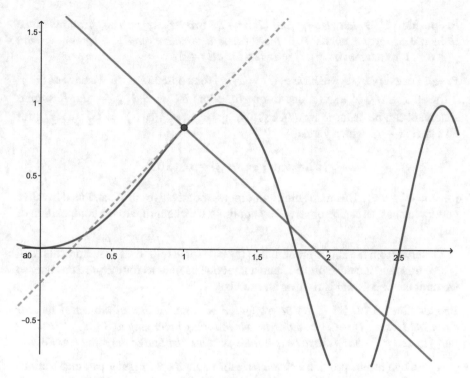

Fig. 1.14 Normal line

Taking derivatives in the previous equality, we get $df(\alpha(t_0)) \cdot d\alpha(t_0) \equiv 0$, which is equivalent to

$$\left(\frac{\partial f}{\partial x}(P), \frac{\partial f}{\partial y}(P) \right) \perp \alpha'(t_0).$$

The tangent line has $\alpha'(t_0)$ as direction vector, so the tangent line of C at P is determined by (1.10) which is perpendicular to the tangent line at P (Fig. 1.14).

Definition 1.3.7 The **normal line** at a point P of a regular curve C is the line at P which is perpendicular to the tangent line at P.

Corollary 1.3.8 *Let $C = \{(x, y) \in \mathbb{R}^2 : f(x, y) = 0\}$ be a regular curve, and $P = (x_0, y_0) \in C$. The normal line of C at P has $\left(\frac{\partial f}{\partial x}(P), \frac{\partial f}{\partial y}(P) \right)$ as its direction vector.*

Given two regular parametrizations of the same arc (I_1, α_1) and (I_2, α_2), there exists a link between them, so that one can obtain all regular parametrizations of certain curve from the knowledge of one of them.

Proposition 1.3.9 *Let (I_1, α_1) and (I_2, α_2) be two regular parametrizations of the same arc of regular curve. Then there exists a bijective map $\gamma : I_1 \rightarrow I_2$ with $\gamma \in \mathcal{C}^\infty(I_1)$ such that $\alpha_2(\gamma(t)) = \alpha_1(t)$, for all $t \in I_1$.*

Proof For every $t \in I_1$ we have $\alpha_1(t) = \alpha_2(s)$ for a unique $s \in I_2$. We have $\alpha_2'(s) \neq 0$ for all $s \in I_2$, so α_2 is a one-to-one function: $\alpha_2^{-1} : \alpha_2(I_2) \rightarrow I_2$. A similar statement can be deduced for (I_1, α_1). We get $s = \alpha_2^{-1}(\alpha_1(t)) = \alpha_2^{-1}(\alpha_2(s))$, for all $t \in I_1, s \in I_2$, which yields

$$\alpha_1(t) = \alpha_2(s) \rightarrow \alpha_1(t) = \alpha_2(\gamma(t)),$$

for $\gamma = \alpha_2^{-1} \circ \alpha_1$. This map admits all orders of derivatives in I_1, and its derivative never vanishes, as can be observed by applying the chain rule to the composition of maps.

Observe from the above proof that $\gamma'(t) \neq 0$ for all $t \in I_1$. The reciprocal result is also true, and it can be proved in an analogous manner as for the previous one, so we omit the details and leave it as an exercise.

Proposition 1.3.10 *Let (I, α) be a regular parametrization of an arc of regular curve, and let $\gamma : I_1 \rightarrow I_2$ be a one-to-one mapping belonging to $\mathcal{C}^\infty(I_1)$. Then, the pair $(I_2, \alpha \circ \gamma^{-1})$ turns out to be a regular parametrization of the same regular arc.*

These two results provide a characterization relating all regular parametrizations of an arc of regular curve. We observe that this relation serves as a different approach in order to prove the independence of the choice of a regular parametrization to determine the tangent line of a regular curve at a point.

Definition 1.3.11 Let (I, α) be a regular parametrization. The **velocity vector** associated to (I, α) is the function $v_\alpha : I \rightarrow \mathbb{R}$ defined by $v_\alpha(t) = \|\alpha'(t)\|$, for all $t \in I$, where $\|\cdot\|$ stands for the Euclidian norm.

The vector $T_\alpha(t) = \alpha'(t)/\|\alpha'(t)\|$ is known as the **tangent unit vector** associated to (I, α) at the point $\alpha(t)$. We define the **normal unit vector** associated to (I, α) at $\alpha(t)$ as the vector obtained by rotating $T_\alpha(t)$ 90 degrees counterclockwise. It is denoted by $N_\alpha(t)$.

It is worth remarking that in the literature we can find different definitions of the normal unit vector. One of them is that of Definition 1.3.11. Another possible choice is the one adopted in Chap. 2 when dealing with spatial curves. There, the normal unit vector is defined in terms of the derivative of the velocity vector. The subsequent elements emerging from this concept may vary depending of the source followed. However, the modification on the construction of the normal vector only varies its orientation, and essentially both provide analogous results. The main reason for the choice made in this present text has a strong advantage, which will be clarified in this chapter. The main disadvantage of the definition adopted here is that it can not be translated to space curves, as the notion of "counterclockwise" can not be generalized in three dimensions.

Observe moreover that given a regular parametrization (I, α), the set

$$\{\alpha(t); \{T_\alpha(t), N_\alpha(t)\}\}$$

is an orthogonal coordinate system in the Euclidean plane $\mathbb{R}^2$, for all $t \in I$.

Due to the fact that the curves are located in the Euclidean plane $\mathbb{R}^2$, one can compute lengths, angles, distances, etc. from the knowledge of the parametrization of a curve.

Proposition 1.3.12 *Let (I, α) be a regular parametrization, and let $t_0 \in I$. For every $t \in I$, the* **arc length** *between $\alpha(t_0)$ and $\alpha(t)$ is given by*

$$\int_{t_0}^t \|\alpha'(u)\| \, du. \tag{1.11}$$

The value of (1.11) does not depend on the choice of the regular parametrization of the arc, so one can say that (1.11) defines the arc length of the regular curve itself.

Outline of the Proof The expression (1.11) is attained by passing to the limit in the Riemann sums of the lengths of lines approximating the curve. Roughly speaking, given $t_0, t \in I$, we consider a partition $\{t_0, t_1, \dots, t_n = t\}$ of the interval $[t_0, t]$ (we proceed analogously for an interval of the form $[t, t_0]$). The sum of the length of all segments joining the points $\alpha(t_j)$ and $\alpha(t_{j+1})$ for $j = 0, \dots, n - 1$ is given by

$$L_n = \sum_{j=0}^{n-1} \sqrt{(\alpha_1(t_j) - \alpha_1(t_{j+1}))^2 + (\alpha_2(t_j) - \alpha_2(t_{j+1}))^2},$$

where we write $\alpha(s) = (\alpha_1(s), \alpha_2(s))$, $s \in I$. It holds that

$$L_n = \sum_{j=0}^{n-1} (t_{j+1} - t_j) \sqrt{\left(\frac{\alpha_1(t_j) - \alpha_1(t_{j+1})}{t_{j+1} - t_j}\right)^2 + \left(\frac{\alpha_2(t_j) - \alpha_2(t_{j+1})}{t_{j+1} - t_j}\right)^2}.$$

This last expression is associated to the Riemann sums of the function $t \mapsto \|\alpha'(t)\|$. We conclude that L_n tends to the expression in (1.11) when n tends to infinity (Fig. 1.15).

The independence with respect to the regular parametrization chosen is a direct consequence of Proposition 1.3.9 and the change of variables in the integral (1.11) given by the change of parameter $\gamma(t)$ relating two regular parametrizations of the arc of curve.

The curvature of a curve at a point measures how much the direction of its tangent line varies when the point under study in the curve is moved slightly. In order to state the concept of curvature of a regular curve at a point in terms of a regular parametrization of the curve, it is important to fix an adequate regular

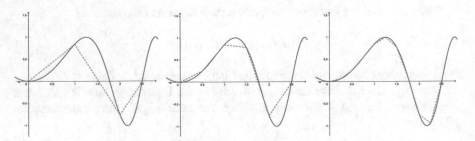

Fig. 1.15 Successive approximations of the length of a curve by segments

parametrization. That parametrization should traverse every point of the curve at the same rate, say 1. For example, the parametrizations

$$\alpha_k : (0, \frac{2\pi}{k}) \to \mathbb{R}^2, \quad \alpha_k(t) = (cos(kt), \sin(kt))$$

are both regular parametrizations of the same arc of curve contained in the unit disc, for all fixed $k > 0$. However, a slight variation of the parameter k would entail large or small variations of the point in the curve. Also, we can consider

$$\alpha_e : \mathbb{R} \to \mathbb{R}^2, \quad \alpha_e(t) = (\cos\left(\frac{2\pi}{1 + \exp(-t)}\right), \sin\left(\frac{2\pi}{1 + \exp(-t)}\right),$$

which is a regular parametrization of the same arc, traversing each point at a different velocity.

Definition 1.3.13 We say that a regular parametrization (I, α) is an **arc length parametrization** (or a **natural parametrization**) if $\left\| \alpha'(t) \right\| = 1$ for every $t \in I$.

Given a regular parametrization, there is always a reparametrization of the same arc, using an arc length parametrization.

Proposition 1.3.14 *Let (I, α) be a regular parametrization. There exists a natural parametrization of $\alpha(I)$.*

Proof Let $\gamma : I \to \gamma(I)$ be defined by

$$\gamma(t) = \int_{t_0}^{t} \left\| \alpha'(u) \right\| du.$$

Then γ is a change of parameter:

- γ is well defined and a one-to-one function in view of the properties of the integral of a continuous and positive function.
- The fundamental theorem of calculus guarantees that $\gamma \in C^{\infty}(I)$. In addition to this, $\gamma'(t) = \left\| \alpha'(t) \right\|$ for all $t \in I$.

We consider $\beta : \gamma(I) \to \mathbb{R}^2$ the function $\beta(t) = \alpha(\gamma^{-1}(t))$. From Proposition 1.3.10 we find that $(\gamma(I), \beta)$ is a regular reparametrization of the same curve. We conclude the proof by verifying that this parametrization is natural. Let $t \in \gamma(I)$. We have

$$\beta'(t) = \alpha'(\gamma^{-1}(t))(\gamma^{-1})'(t) = \alpha'(\gamma^{-1}(t))\frac{1}{\gamma'(\gamma^{-1}(t))} = \alpha'(\gamma^{-1}(t))\frac{1}{\|\alpha'(\gamma^{-1}(t))\|}.$$

This means that $\|\beta'(t)\| = 1$ for all $t \in \gamma(I)$.

Observe that the above proof allows us to construct, at least theoretically, a natural parametrization associated to a regular one.

Proposition 1.3.15 *Let (I_1, β_1) and (I_1, β_2) be two natural parametrizations of the same arc. Then there exists $c \in \mathbb{R}$ such that $\beta_1(t) = \beta_2(\pm t + c)$ for all $t \in I_1$, and the intervals are related accordingly.*

Proof Regarding Proposition 1.3.9, there exists $\gamma : I_1 \to I_2$ such that $\beta_1(t) = \beta_2(\gamma(t))$, $t \in I_1$. Taking derivatives in the last expression yields $\beta_1'(t) = \gamma'(t)\beta_2'(\gamma(t))$. Taking norms in the previous equality, and bearing in mind that both parametrizations are natural, yields $1 = |\gamma'(t)|$. The fact that $\gamma \in C^\infty(I_1)$ means $\gamma(t) = \pm \int 1 dt = \pm t + c$, for some $c \in \mathbb{R}$.

We now state the definition of curvature of a regular curve at a point. First, this is done for curves described by a natural parametrization; we then generalize that definition for any other regular parametrization.

Definition 1.3.16 Let (I, α) be a natural parametrization. We consider the function $\kappa_\alpha : I \to \mathbb{R}$ given by

$$T_\alpha'(t) = \alpha''(t) = \kappa_\alpha(t)N_\alpha(t). \tag{1.12}$$

We say that $\kappa_\alpha(t)$ is the **curvature** of (I, α) at $t \in I$.

In addition to that, $\alpha(t_0) \in \alpha(I)$ is an **inflection point** of the curve $\alpha(I)$ if $\kappa_\alpha(t_0) = 0$, or equivalently, if $\alpha''(t_0) = 0$.

We observe that (1.12) makes sense and defines a unique value of $\kappa_\alpha(t)$ because the vector $\alpha''(t)$ is orthogonal to $\alpha'(t)$, thus it is proportional to $N_\alpha(t)$.

Moreover, the previous definition does not depend essentially on the natural parametrization chosen. Given two natural parametrizations (I_1, α_1) and (I_2, α_2) of the same curve C, then $\alpha_1 \circ \gamma = \alpha_2$ for some bijection $\gamma : I_2 \to I_1$, by Proposition 1.3.9. Taking norms in the previous equality yields $|\gamma'(t)| = 1$ for all $t \in I_1$. Because γ' is a continuous function, it has to be $\gamma(t) = \pm t + c$, for some $c \in \mathbb{R}$, and the curvature only varies in its when changing the natural parametrization.

The remark after Definition 1.3.11 can be stated at this point. Under the definition of normal vector that we have adopted, the curvature can be positive or negative. This results in a geometric meaning which can be explained as follows.

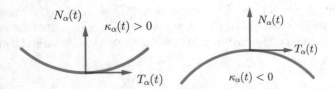

Fig. 1.16 Curvature at a point, I

Example 1.3.17 Let us consider the unit circle, and the natural parametrizations $((0, 2\pi), \alpha_1)$ and $((0, 2\pi), \alpha_2)$ defined by

$$\alpha_1(t) = (\cos(t), \sin(t)), \quad \alpha_2(t) = (\cos(t), -\sin(t)), \quad t \in (0, 2\pi),$$

of an arc contained in the unit circle. Referring to Definition 1.3.16, we get that $\kappa_{\alpha_1}(t) = -1$ and $\kappa_{\alpha_2}(t) = 1$ for all $t \in (0, 2\pi)$. Observe the normal vector points towards the center of the circle when the curvature is positive, and away from the center of the circle otherwise. Figure 1.16 illustrates the situation.

Let us consider a general situation and take a natural parametrization of a curve (I, α). Choose $P = \alpha(t) \in \alpha(I)$ and assume that $\kappa_\alpha(t) \neq 0$. Then one can draw the tangent circle to the curve at P which better fits the curve. Proceeding in the same way for every point on the curve with curvature different from zero, one can merge the points into two categories, regarding the curvature. An sketch of the situation is shown in Fig. 1.17. An illustrating example can be found in the QR code of Fig. 1.17.

We proceed to extend the notion of curvature for regular curves. The key points are Propositions 1.3.9 and 1.3.14: given a regular parametrization of an arc, the value of the curvature at each point should correspond (in absolute value) to that of a natural reparametrization of the curve. The different choices of a natural reparametrization causes a change of sign in the curvature, leading to an up-to-sign coherent definition for the curvature of a regular parametrization.

Definition 1.3.18 Let (I_1, α) be a regular parametrization. We define $\kappa_\alpha(t) := \kappa_\beta(\gamma(t))$, for every $t \in I_1$, and $\alpha = \beta \circ \gamma$, (I_2, β) being a natural parametrization of $\alpha(I_1)$, and where $\gamma'(t) > 0$.

Example 1.3.19 Let $((0, 2\pi), \alpha)$ be a parametrization of an arc of circle of radius $R > 0$ given by

$$\alpha(t) = (R \cos(t), R \sin(t)), \quad t \in (0, 2\pi).$$

Regarding Proposition 1.3.14, the change of parameter

$$\gamma(t) = \int_0^t \sqrt{(-R \sin(t))^2 + (R \cos(t))^2} \, dt = \int_0^t R \, dt = Rt$$

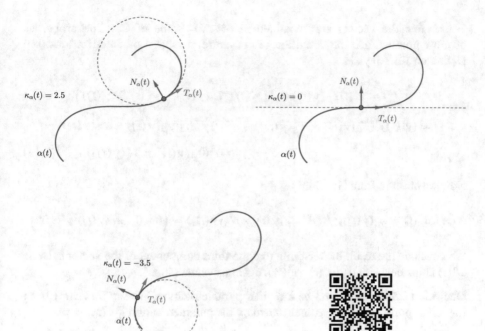

Fig. 1.17 Curvature at a point, II. QR Code 3

provides a natural parametrization of the arc of circle, $\beta = \alpha \circ \gamma^{-1}$. We have

$$\beta(s) = (R\cos(\frac{1}{R}s),\ R\sin(\frac{1}{R}s)), \quad s \in (0, 2\pi R).$$

It holds that $\beta''(s) = \kappa_\beta(s)N_\beta(s)$, for all $s \in (0, 2\pi R)$, which yields $\kappa_\beta(\gamma(t)) = \kappa_\alpha(t) = 1/R$.

Proposition 1.3.20 *Let (I, α) be a regular parametrization. It holds that*

$$\kappa_\alpha(t) = \frac{det(\alpha'(t), \alpha''(t))}{\|\alpha'(t)\|^3}, \quad t \in I.$$

Proof Let (I_2, β) be a natural parametrization of $\alpha(I_1)$. We consider the change of parameter $\gamma : I \to I_2$ with $\alpha = \beta \circ \gamma$, such that $\gamma'(t) > 0$ for all $t \in I$. One has

$$\alpha'(t) = \beta'(\gamma(t))\gamma'(t) = T_\beta(\gamma(t))\gamma'(t), \quad t \in I,$$

$$\alpha''(t) = T'_\beta(\gamma(t))(\gamma'(t))^2 + T_\beta(\gamma(t))\gamma''(t), \quad t \in I.$$

We assume the vectors are contained in the XY plane of $\mathbb{R}^3$. The properties of cross product, and the definition of curvature in natural parametrizations (see Definition 1.3.16) yield

$$\alpha'(t) \times \alpha''(t) = \gamma'(t)T_\beta(\gamma(t)) \times \left((\gamma'(t))^2 T'_\beta(\gamma(t)) + \gamma''(t)T_\beta(\gamma(t))\right)$$

$$= (\gamma'(t))^3 (T_\beta(\gamma(t)) \times T'_\beta(\gamma(t))) + \gamma'(t)\gamma''(t)(T_\beta(\gamma(t)) \times T_\beta(\gamma(t)))$$

$$= (\gamma'(t))^3 (T_\beta(\gamma(t)) \times T'_\beta(\gamma(t))). \qquad (1.13)$$

Now we observe from (1.12) that

$$\alpha'(t) \times \alpha''(t) = \kappa_\beta(\gamma(t))(\gamma'(t))^3 (T_\beta(\gamma(t)) \times N_\beta(\gamma(t))) = \begin{pmatrix} 0 & 0 & \kappa_\beta(\gamma(t))(\gamma'(t))^3 \end{pmatrix}.$$

We conclude the result by verifying that the third component of the vector $\alpha'(t) \times \alpha''(t)$ coincides with $\det(\alpha'(t), \alpha''(t))$, again as vectors in $\mathbb{R}^3$.

Definition 1.3.21 Let (I, α) be a regular parametrization. We say that $\alpha(t_0)$ is an **inflection point** of the curve determined by the parametrization if $\kappa(t_0) = 0$.

This section concludes with a result which states that all the relevant information of a parametric curve is stored in its curvature. A more general result is stated in Theorem 2.2.37, so we omit the details on the proof at this stage.

Theorem 1.3.22 (Fundamental Theorem of Plane Curves) *Let $\kappa : I \to \mathbb{R}$ belong to $C^\infty(I)$, where $\emptyset \neq I \subseteq \mathbb{R}$ is an open interval. Then there exists a natural parametrization (I, α) such that $\kappa(t) = \kappa_\alpha(t)$, $t \in I$. Any other regular curve under this property results from a rigid transformation of the first curve in $\mathbb{R}^2$ (i.e., composition of translations, stiff rotations and/or reflections).*

1.4   Conics

There are different approaches to conics. These objects appear as the sections of a cone cut by a plane in different positions (see Figs. 1.18 and 1.19). We also refer to Hanh (2012) for an in-depth study of conics from this point of view.

Another approach to conics is to consider them as the locus

- of the points in a plane such that the sum of the distances to two fixed points (foci) is the same constant (**ellipse**).
- of the points in a plane such that the difference of the distances to two fixed points (foci) is the same fixed constant (**hyperbola**).
- of the points in a plane which are equidistant to a fixed point (focus) and a fixed line (directrix) (**parabola**).

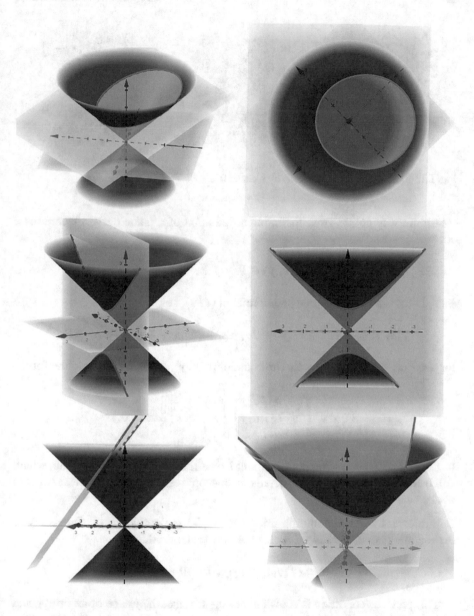

Fig. 1.18 Conics as sections of a cone by a plane, I

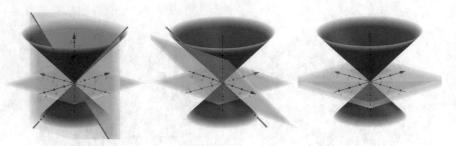

Fig. 1.19 Conics as sections of a cone by a plane, II

A third way comes from the implicit definition of a plane curve. A conic is a curve

$$C = \{(x, y) \in \mathbb{R}^2 : f(x, y) = 0\},$$

such that f is a second degree polynomial in (x, y), i.e.,

$$f(x, y) = a_{11}x^2 + a_{22}y^2 + 2a_{12}xy + 2a_{01}x + 2a_{02}y + a_{00} = 0,$$

for some $a_{ij} \in \mathbb{R}$, $0 \leq i, j \leq 2$. The equation $f(x, y) = 0$ is rewritten in the form

$$\begin{pmatrix} 1 & x & y \end{pmatrix} \begin{pmatrix} a_{00} & a_{01} & a_{02} \\ a_{01} & a_{11} & a_{12} \\ a_{02} & a_{12} & a_{22} \end{pmatrix} \begin{pmatrix} 1 \\ x \\ y \end{pmatrix} = x^T M x = 0.$$

Let us recall some general facts about isometries in the Euclidean plane, which will be used in the sequel. More details on this topic can be found in Linear Algebra books such as Lang (1986).

Definition 1.4.1 An endomorphism $f : \mathbb{R}^2 \to \mathbb{R}^2$ in the Euclidean plane $\mathbb{R}^2$ is an **orthogonal transformation** if it preserves the scalar product:

$$\langle v_1, v_2 \rangle = \langle f(v_1), f(v_2) \rangle, \text{ for all } v_1, v_2 \in \mathbb{R}^2.$$

This previous condition is equivalent to the fact that the image of an orthogonal basis is an orthonormal basis. These transformations are reduced in the Euclidean plane to rotations (direct transformations) and orthogonal symmetries (inverse transformations).

Two affine bases of the real plane $\{O, \{e_1, e_2\}\}$ and $\{O', \{u_1, u_2\}\}$, for some $O, O' \in \mathbb{R}^2$ and $\{e_1, e_2\}$, $\{u_1, u_2\}$ being orthogonal bases of $\mathbb{R}^2$, are related by a translation of vector $\overrightarrow{OO'}$, and an orthogonal transformation (rotation and/or symmetry) which convert the first basis of $\mathbb{R}^2$ into the other (see Fig. 1.20).

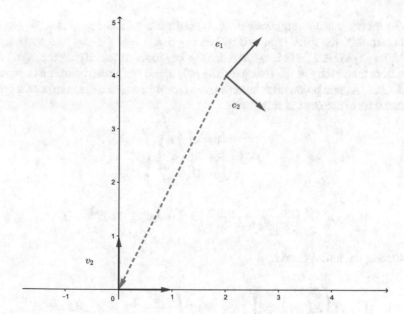

Fig. 1.20 Orthogonal transformation of a coordinate system

We now give some details on the classification of conics in terms of some of the elements appearing in the matrix defining the conic, which turn out to be invariant by translations or orthogonal transformations. Through an inspection of the result of such transformations made on a general conic, we determine every (essentially) different conic.

Let M be the matrix of a conic in the canonical coordinate system of $\mathbb{R}^2$, and let M' be the matrix of that conic in a new orthogonal coordinate system. The associated matrix of change of coordinates is given by

$$Q = \left(\begin{array}{c|c} 1 & O \\ \hline c & Q_0 \end{array} \right),$$

where $c^T = (c_1, c_2)$ stands for the coordinates of the origin of coordinates in the new reference, and Q_0 is the orthogonal matrix associated to the change of basis in the corresponding vector space. We write

$$M = \left(\begin{array}{c|c} d & b^T \\ \hline b & M_0 \end{array} \right),$$

with $d \in \mathbb{R}$, and $b \in \mathcal{M}_{2 \times 1}(\mathbb{R})$. The spectral theorem allows us to choose Q_0 in such a way that $M'_0 = Q_0^T M_0 Q_0$ is a diagonal matrix, say $M'_0 = \mathrm{diag}(\lambda_1, \lambda_2)$, for some $\lambda_1, \lambda_2 \in \mathbb{R}$.

We observe that the eigenvalues of M_0 coincide with those of M_0': λ_1, λ_2. In addition to this, the trace is invariant, i.e., $a_{11} + a_{22} = \lambda_1 + \lambda_2$, and $\det(M_0) = \det(M_0') = \lambda_1\lambda_2$. All these invariants are the key tools for clasifying the conic and come from the unicity of the characteristic polynomial of a matrix after a change of coordinates. At this point, after an orthogonal transformation, the matrix of a conic in general form displays as follows:

$$
\begin{pmatrix} 1 & x^{\star\star} & y^{\star\star} \end{pmatrix} \begin{pmatrix} b_{00} & b_{01} & b_{02} \\ b_{01} & \lambda_1 & 0 \\ b_{02} & 0 & \lambda_2 \end{pmatrix} \begin{pmatrix} 1 \\ x^{\star\star} \\ y^{\star\star} \end{pmatrix}
$$

$$
= (x^{\star\star})^T \left(\begin{array}{c|c} d' & b'^{T} \\ \hline b' & \mathrm{diag}(\lambda_1, \lambda_2) \end{array} \right) x^{\star\star} = (x^{\star\star})^T M' x^{\star\star} = 0,
$$

substituting the initial matrix

$$
\begin{pmatrix} 1 & x & y \end{pmatrix} \begin{pmatrix} a_{00} & a_{01} & a_{02} \\ a_{01} & a_{11} & a_{12} \\ a_{02} & a_{12} & a_{22} \end{pmatrix} \begin{pmatrix} 1 \\ x \\ y \end{pmatrix} = x^T \left(\begin{array}{c|c} d & b^T \\ \hline b & M_0 \end{array} \right) x = x^T M x = 0.
$$

Definition 1.4.2 The conics described by a matrix M with $\det(M) = 0$ are known as **reducible conics**. Otherwise, we say that the conic is an **irreducible conic** .

We observe from the previous arguments that the determinant of the matrix defining the conic does not depend on the coordinate system considered.

Proposition 1.4.3 *Every reducible conic is a pair of lines (complex or real).*

Proof Given a pair of real or complex lines, we have that the conic defined by them is associated to the equation $f(x, y) = 0$, where

$$
f(x, y) = (b_0 + b_1 x + b_2 y)(c_0 + c_1 x + c_2 y), \tag{1.14}
$$

for some $x, y \in \mathbb{R}$ or $\mathbb{C}$. It is straight to check that the corresponding matrix determined by this conic satisfies the condition that its determinant is null. Reciprocally, given a conic represented by a matrix M with $\det(M) = 0$, one can write its defining equation as a product of two lines by comparing coefficients in the form of a factorization of the form (1.14).

Let λ_1, λ_2 be the eigenvalues of M, and assume that one has proceeded to perform the mentioned orthogonal transformation. In the previous situation, one has:

• If $\lambda_1 \lambda_2 \neq 0$, then the translation

$$
(x, y) \mapsto (x - \frac{b_{01}}{\lambda_1}, y - \frac{b_{02}}{\lambda_2})
$$

transforms the initial conic into that of equation

$$\lambda_1 x^2 + \lambda_2 y^2 = -\frac{\det(M)}{\lambda_1 \lambda_2}, \tag{1.15}$$

which is an **ellipse** or an **hyperbola**, depending on the signs of λ_1, λ_2 and $\det(M)$.

- If $\lambda_1 \lambda_2 = 0$, and $\lambda_2 \neq 0$ (the same argument holds in the symmetric situation), then one has:

  - If $\det(M) \neq 0$, the translation

$$(x, y) \mapsto \left(x - \frac{b_{00}\lambda_2 - b_{02}^2}{2\lambda_2 b_{01}}, \ y - \frac{b_{02}}{\lambda_2} \right)$$

  determines the conic

$$y^2 = -2\sqrt{-\frac{\det(M)}{\lambda_2^3}} x, \tag{1.16}$$

  which is a **parabola**.
  - If $\det(M) = 0$, we get a pair of **parallel lines**.

More precisely, let M_0 be the submatrix of M formed with the quadratic terms of the conic and let λ_1, λ_2 be its eigenvalues. Then, one has the following classification of a conic. The whole classification directly follows from the statements above, and Proposition 1.4.3. In the following, the **signature** (a, b) of a matrix signifies that such matrix has a positive eigenvalues and b negative eigenvalues.

Classification of Conics

$\det(M) \neq 0$. Irreducible conic

1. $\det(M_0) \neq 0$.

   1.1. $\det(M_0) > 0$. Ellipse

       1.1.1. $\text{sign}(\det(M)) = \text{sign}(a_{11} + a_{22})$. **IMAGINARY ELLIPSE**
       1.1.2. $\text{sign}(\det(M)) \neq \text{sign}(a_{11} + a_{22})$. **REAL ELLIPSE**

   1.2. $\det(M_0) < 0$. **HYPERBOLA**

2. $\det(M_0) = 0$. **PARABOLA**

$\det(M) = 0$. Reducible conic

3. $\det(M_0) \neq 0$. Non-parallel lines

   3.1. $\det(M_0) > 0$. **PAIR OF IMAGINARY NON-PARALLEL LINES**
   3.2. $\det(M_0) < 0$. **PAIR OF REAL NON-PARALLEL LINES**

4. $\det(M_0) = 0$. Real or complex parallel or coincident lines.

    4.1. $\lambda_1 = \lambda_2 = 0$. There are not quadratic terms. **REAL LINE**.

    4.2. $\lambda_1\lambda_2 = 0$.

        4.2.1. Signature of M is $(1, 1)$, **PARALLEL REAL LINES**.

        4.2.2. Signature of M is $(2, 0)$ or $(0, 2)$, **IMAGINARY PARALLEL LINES**.

        4.2.3. $\operatorname{rank}(M) = 1$. **COINCIDENT LINES**.

From the above discussion, every implicit equation of a conic arises. For practical purposes, we will only make use of the non-imaginary ones. We leave to the reader to develop the details of the proof of the above classification in the case of reducible conics. In view of Proposition 1.4.3 one can provide more details on the coefficients b_j, c_j for $0 \leq j \leq 2$ in (1.14) in each situation.

Proposition 1.4.4 *After an orthogonal change of coordinates, the following statements correspond to the equations of each conic:*

- *Real and imaginary ellipse.*

  - *Real ellipse: $\frac{x^2}{a^2} + \frac{y^2}{b^2} = 1$, for some $a, b \in \mathbb{R}^\star$.*
  - *Imaginary ellipse: $\frac{x^2}{a^2} + \frac{y^2}{b^2} = -1$, for some $a, b \in \mathbb{R}^\star$.*

- *Hyperbola: $\frac{x^2}{a^2} - \frac{y^2}{b^2} = 1$, for some $a, b \in \mathbb{R}^\star$.*
- *Parabola: $y^2 = a^2x$, for some $a \in \mathbb{R}^\star$.*
- *Pair of non-parallel lines: $(a_1x + b_1y)(a_2x + b_2y) = 0$, for $a_1, b_1, a_2, b_2 \in \mathbb{C}$, with the vector (a_1, b_1) not being proportional to (a_2, b_2). The secant lines are real if $a_1, a_2, b_1, b_2 \in \mathbb{R}$.*
- *Parallel lines: $(ax + by)(ax + by + c) = 0$, for $a, b, c \in \mathbb{C}$, with $c \neq 0$. The secant lines are real if $a, b, c \in \mathbb{R}$.*
- *Coincident lines: $(ax + by)^2 = 0$, for $a, b \in \mathbb{R}$.*
- *Real line: $ax + by + c = 0$, for $a, b, c \in \mathbb{R}$.*

Proof The result is a direct consequence of (1.15), (1.16), and the classification of the reducible conics given above.

Example 1.4.5 Let

$$C_1 = \{(x, y) \in \mathbb{R}^2 : 2x^2 - 2y^2 + 2xy + y - 4 = 0\}.$$

It holds that the matrix

$$M_1 = \begin{pmatrix} -4 & 0 & 1/2 \\ 0 & 2 & 1 \\ 1/2 & 1 & -2 \end{pmatrix}$$

determines the associated matrix to C_1. It holds that C_1 is a hyperbola because $\det(M_1) = 39/2$ and $\det(M_1)_0 = -5$.

Let

$$C_2 = \{(x, y) \in \mathbb{R}^2 : x^2 + y^2 + 2xy + y - 4 = 0\}.$$

It holds that the matrix

$$M_2 = \begin{pmatrix} -4 & 0 & 1/2 \\ 0 & 1 & 1 \\ 1/2 & 1 & 1 \end{pmatrix}$$

determines the associated matrix to C_2. It holds that C_2 determines a parabola because $\det(M_2) = -1/4$ and $\det(M_2)_0 = 0$.

Let

$$C_3 = \{(x, y) \in \mathbb{R}^2 : x^2 + 2y^2 + 3xy + 2x + 3y + 1 = 0\}.$$

It holds that the matrix

$$M_3 = \begin{pmatrix} 1 & 1 & 3/2 \\ 1 & 1 & 3/2 \\ 3/2 & 3/2 & 2 \end{pmatrix}$$

determines the associated matrix to C_3. It holds that C_3 determines a pair of secant lines because $\det(M_3) = 0$ and $\det(M_3)_0 < 0$.

In order to determine both lines, one could try to write the conic in the form $(ax + by + c)(dx + ey + f) = 0$ and compare coefficients. Several degrees of freedom can be found on the way.

Let

$$C_4 = \{(x, y) \in \mathbb{R}^2 : x^2 + 3y^2 + 3xy + 2x + 3y + 1 = 0\}.$$

It holds that the matrix

$$M_4 = \begin{pmatrix} 1 & 1 & 3/2 \\ 1 & 1 & 3/2 \\ 3/2 & 3/2 & 3 \end{pmatrix}$$

determines the associated matrix to C_4. It holds that C_4 determines a pair of imaginary secant lines (a single point in $\mathbb{R}^2$, indeed C_4 is reduced to $(-1, 0)$) because $\det(M_4) = 0$ and $\det(M_3)_0 > 0$.

Some elements of the conics are straightforward to find when considering their matrix representation, such as the existence of one or more centers, the existence and determination of the axis, etc.

Definition 1.4.6 A point in $\mathbb{R}^2$ is a **center** of a conic if it is a center of symmetry of the conic. The **axes** of a conic with center are the symmetry axis of such conic.

Proposition 1.4.7 *Given a conic of associated matrix*

$$M = \left(\frac{d \mid b^T}{b \mid M_0} \right),$$

then the conic admits a center if and only if $rank(M_0) = rank(M_0 \mid b)$. *Moreover, the coordinates of all centers of the conic are given by the solutions of the system*

$$M_0 \begin{pmatrix} x \\ y \end{pmatrix} = -b.$$

Proof Let $C = (c_1, c_2) \in \mathbb{R}^2$. We consider a change of coordinates which locates the origin of its coordinates at C. Then the linear terms of the new equation of the conic are given by $M(c_1 c_2)^T + b$. It is a necessary and sufficient condition for the conic to have C as a center that the equation does not vary when x is replaced by $-x$ and y is replaced by $-y$. This means that the linear terms vanish.

In view of the previous result, the conics with center of symmetry are the following:

- One center $\Rightarrow$ ellipse and hyperbola, pair of non-parallel lines.
- One line of centers $\Rightarrow$ a pair of coincident or parallel lines.

The case of one line does not have quadratic terms. In practice, it could be treated as coincident lines.

Proposition 1.4.8 *Let* $C = (c_1, c_2)$ *be a center of symmetry of a conic. Then the associated axis of the conic are lines at* C *with directions given by the eigenvectors of the submatrix of quadratic terms.*

Proof We recall that every linear term of the matrix of the conic obtained after translating the center of coordinates to C has to vanish. The submatrix of quadratic terms, M_0, is symmetric, so the spectral theorem guarantees the existence of an orthogonal basis of $\mathbb{R}^2$ of eigenvectors of M_0. Let $Q_0 \in \mathcal{M}_{2\times 2}(\mathbb{R})$ be the matrix with columns given by the eigenvectors of the orthonormal basis constructed. The change of coordinates associated to the matrix

$$Q = \left(\frac{1 \mid O}{0 \mid Q_0} \right)$$

transforms the matrix of the conic into

$$M' = \left(\begin{array}{c|c} d & O \\ \hline 0 & M'_0 \end{array}\right),$$

with $M'_0 = \mathrm{diag}(\lambda_1, \lambda_2)$ and some $d \in \mathbb{R}$. The equation of the conic in such coordinates is

$$\lambda_1 (x^{\star\star})^2 + \lambda_2 (y^{\star\star})^2 + d = 0,$$

so the axes of symmetry of the conic are $x^{\star\star} = 0$ and $y^{\star\star} = 0$. The inverse changes lead to the lines defined by

$$(x, y) = (c_1, c_2) + tu_1, \quad t \in \mathbb{R},$$

$$(x, y) = (c_1, c_2) + tu_2, \quad t \in \mathbb{R},$$

in the initial system of coordinates. Here $\{u_1, u_2\}$ stands for the orthogonal basis of eigenvectors of $\mathbb{R}^2$ with respect to Q_0.

A different argument can be followed to determine the asymptotes of a hyperbola.

Definition 1.4.9 An **asymptote** of a hyperbola is a line passing through the center of the hyperbola, and at zero distance to the hyperbola, but without points in common with it.

Proposition 1.4.10 *Every hyperbola admits two asymptotes. The equations of the asymptotes of a hyperbola are given by*

$$y - c_2 = m(x - c_1),$$

where $C = (c_1, c_2)$ is the center of the hyperbola, and $m \in \mathbb{R}$ is a solution of

$$\begin{pmatrix} 1 & m \end{pmatrix} M_0 \begin{pmatrix} 1 \\ m \end{pmatrix} = 0. \tag{1.17}$$

Proof Assume the origin of coordinates is the center of the hyperbola without loss of generality. The matrix of the conic is of the form

$$M = \left(\begin{array}{c|c} c & O \\ \hline 0 & M_0 \end{array}\right), \qquad M_0 = \begin{pmatrix} a_{11} & a_{12} \\ a_{12} & a_{22} \end{pmatrix}.$$

In order for $y = mx$ to be an asymptote, it has to satisfy the equation of the hyperbola for $x \to \infty$, which yields that

$$a_{22}(mx)^2 + 2a_{12}mx^2 + a_{11}x^2 + c = 0$$

at the limit $x \to \infty$, that is,

$$m^2 a_{22} + 2a_{12}m + a_{11} + \underbrace{\frac{c}{x^2}}_{\to 0} = 0,$$

or equivalently,

$$(1 \ m) \ M_0 \begin{pmatrix} 1 \\ m \end{pmatrix} = 0.$$

Observe that the system in (1.17) determines a quadratic equation in m, with real roots due to its discriminant is positive, because the conic is a hyperbola.

Observe that if $a_{22} = 0$ in the matrix representing a hyperbola, then one only attains one of the asymptotes of the hyperbola with the formula (1.17). That asymptote is a horizontal line. There is a second asymptote passing through the center, and of vertical direction.

So far, we have considered and studied the conics by means of their implicit representation. However, as mentioned above, parametrizations are of great importance when working with a curve. In the next paragraphs, we provide parametrizations describing each real conic.

Algorithm (Parametrization of a Conic)
Input: The implicit equation of a conic, $f(x, y) = 0$.
  Output: A parametrization of the conic.

1. Compute the matrix representation of the conic.
2. Classify the conic, regarding the invariants of the conic.
3. Make appropriate changes of coordinates as described above (translations and isometries). The equation of the conic in the new system of coordinates is determined in Proposition 1.4.4.
4. Distinguish the following cases to provide a parametrization of the conic:

- Real ellipse $\left(\frac{x^2}{a^2} + \frac{y^2}{b^2} = 1 \right)$ can be parametrized by

$$\begin{cases} x(t) = a \cos(t) \\ y(t) = b \sin(t), \quad t \in \mathbb{R}. \end{cases}$$

  Observe that the whole curve except for one point is covered for every interval of length 2π of the parameter.
- Hyperbola $\left(\frac{x^2}{a^2} - \frac{y^2}{b^2} = 1 \right)$ can be parametrized by

$$\begin{cases} x(t) = \pm a \cosh(t) \\ y(t) = b \sinh(t), \quad t \in \mathbb{R}. \end{cases}$$

  Each choice for the sign $+$ or $-$ in the previous parametrization draws one of the branches of the hyperbola.

- Parabola $\left(y^2 = a^2 x\right)$, is parametrized by

$$\begin{cases} x(t) = \frac{t^2}{a^2} \\ y(t) = t, \quad t \in \mathbb{R}. \end{cases}$$

- In case the conic is reducible, each of the lines $ax + by + c = 0$ determining the conic is parametrized as follows:

$$\begin{cases} x(t) = t \\ y(t) = -\frac{c}{b} - \frac{a}{b}t, \quad t \in \mathbb{R} \end{cases} \quad \text{(if } b \neq 0\text{)}$$

$$\begin{cases} x(t) = t - \frac{c}{a} - \frac{b}{a}t \\ y(t) = t, \quad t \in \mathbb{R} \end{cases} \quad \text{(if } a \neq 0\text{)}$$

5. Undo the change of coordinates applied at Step 3.

Corollary 1.4.11 *Every irreducible conic admits a parametrization such that each of its components can be parametrized by:*

- *a linear combination of* $\sin(t)$ *and* $\cos(t)$ *in the case of a real ellipse.*
- *a linear combination of* $\sinh(t)$ *and* $\cosh(t)$ *in the case of a hyperbola.*
- *a second degree polynomial in the case of a parabola.*

Example 1.4.12 Let us consider the conic

$$C = \{(x, y) \in \mathbb{R}^2 : 3x^2 + 3y^2 - 2xy - 8x + 8y + 6 = 0\}.$$

Its associated matrix is given by

$$M = \begin{pmatrix} 6 & -4 & 4 \\ -4 & 3 & -1 \\ 4 & -1 & 3 \end{pmatrix}.$$

This curve is an ellipse. The eigenvalues of the matrix

$$M_0 = \begin{pmatrix} 3 & -1 \\ -1 & 3 \end{pmatrix}$$

are $\lambda_1 = 2$, and $\lambda_2 = 4$. An orthonormal basis of $\mathbb{R}^2$ associated to such eigenvalues is $\{(\frac{1}{\sqrt{2}}, \frac{1}{\sqrt{2}}), (\frac{1}{\sqrt{2}}, \frac{-1}{\sqrt{2}})\}$. The center of the ellipse is the point $P = (1, -1)$. Moving the origin of coordinates to that point, and the coordinate axis to the axis of the ellipse, one arrives at the equation of the ellipse in a different coordinate system:

$$(x^{\star\star})^2 + 2(y^{\star\star})^2 = 1.$$

The previous implicit equation of the ellipse determines the parametrization

$$x^{\star\star}(t) = \cos(t)$$

$$y^{\star\star}(t) = \frac{1}{\sqrt{2}} \sin(t), \quad t \in (0, 2\pi). \tag{1.18}$$

The change of coordinates performed is associated to the rigid transformation

$$\begin{pmatrix} 1 \\ x \\ y \end{pmatrix} = \begin{pmatrix} 1 & 0 & 0 \\ 1 & \frac{1}{\sqrt{2}} & \frac{1}{\sqrt{2}} \\ -1 & \frac{1}{\sqrt{2}} & -\frac{1}{\sqrt{2}} \end{pmatrix} \begin{pmatrix} 1 \\ x^{\star\star} \\ y^{\star\star} \end{pmatrix}. \tag{1.19}$$

Bearing in mind (1.18) and (1.19), we arrive at a parametrization of the ellipse in the original coordinate system given by

$$x(t) = 1 + \frac{1}{\sqrt{2}} \cos(t) + \frac{1}{2} \sin(t)$$

$$y(t) = -1 + \frac{1}{\sqrt{2}} \cos(t) - \frac{1}{2} \sin(t), \quad t \in (0, 2\pi).$$

1.5   Some Conics in Architecture

This section shows different classic plane curves appearing in architectural elements, with the focus on conics. We have decided to provide a small representation of such applications, which remain constant thoughout history.

Ellipse
Ellipses have appeared recursively in art and architecture. An etching by Étienne Dupérac from 1568 shows Campidoglio square in Rome. The pavement of such square, designed by Michelangelo, is based on the figure of an ellipse (see Fig. 1.21), with an statue located at the center of that curve.Although initially designed by Michelangelo, it was only finally realised in 1948. The design is not a true ellipse, as it is composed of straight lines.[1]

Ellipses can also be found in other architectonic elements such as St. Peter's square by Bernini in Rome. Two fountains are located at the two foci of the ellipse (see Fig. 1.22). We also refer to Hanh (2012) for some details on this square from the geometric point of view.

Many other architectural representations make use of ellipses, such as amphitheatres like that in Merida, the floor plan of Lipstick building by Philip Johnson in

---

[1] We thank Kim Williams for completing this information.

Fig. 1.21 Campidoglio square, Rome. Engraving: Étienne Dupérac, 1568

Fig. 1.22 Aerial view of St. Peter's square, in Rome

New York, stadiums like the Maracanã in Rio de Janeiro, Brazil, certain zones of
the El Retiro park in Madrid, etc. The work (Kimberling 2004) is devoted to the
study of ellipses in Washington D.C.

Hyperbola

The stability properties of certain surfaces based on the hyperbola (see Sect. 4.1
and Example 4.4.3) make it appear frequently in buildings: a section by a plane of
the cooling tower of a nuclear reactor draws such figure; the so-called hyperbolic
towers such as Canton Tower in Guangzhou, Zhongyuan Tower in Zhengzhou, etc;
some ship masts; and many other structures based on surfaces formed by hyperbolas.
Usually, they appear as sections in certain buldings and will be studied in more detail
later in Sect. 3.4. Figure 1.23 displays the Cathedral of Brasilia, by Oscar Niemeyer,
whose structure contains hyperbolas inside planes orthogonal to the plan floor.

Fig. 1.23   Cathedral of Brasilia, by Oscar Niemeyer

Fig. 1.24 Oceanogràfic by Félix Candela

Parabola

Much like hyperbolas, parabolas form part of the structure of many buildings. They can be found with appropriate sections of some quadric-like buildings, as it will be pointed out later. Examples of such buildings are Lee Valley VeloPark in London by Hopkins Architects, Warszawa Ochota railway station in Warsaw designed by Arseniusz Romanowicz and Piotr Szymaniak, the Philips Pavilion by Le Corbusier and Iannis Xenakis, etc.

In this direction, we refer to the webpage (Wikipedia 2019b), where both hyperbolic and parabolic structures appear in a long list of hyperboloid structures. Figure 1.24 shows Oceanogràfic in Valencia, Spain, by architect Félix Candela and the structural engineers Alberto Domingo and Carlos Lázaro, whose structure is based on a surface containing parabolas.

1.6   On the Implicitation and Parametrization of Curves

In Sect. 1.1, we stated two different ways to define a curve, either via the points in $\mathbb{R}^2$ which satisfy an implicit equation $f(x, y) = 0$, or by parametrizations. Both are closely related, passing from one to the other (see Theorems 1.1.7 and 1.1.8). Given a regular curve determined by its implicit form, the implicit function theorem guarantees that, locally, there exists a parametrization of the curve. On the other hand, one of the components of a regular parametrization determines an injective map, allowing to construct a function f in two variables such that the points (x, y) satisfying $f(x, y) = 0$ determine the curve locally.

However, the two previous constructions are local, and also some hard problems may appear. Given an implicit equation $f(x, y) = 0$, determining the function $y = y(x)$ or $x = x(y)$ with $f(x, y(x)) = 0$ or $f(x(y), y) = 0$ near a point might be impossible. Given a parametrization, the same holds when searching for the explicit inverse of a function locally.

In recent decades, much attention has been devoted to the so-called **rational algebraic curves**, giving rise to theories and algorithms in this direction, which are efficient in practice and are also capable of treating the curve under a point of view that is not local. Here we are not going into details on this vast and very interesting theory; we will only focus on some direct applications in order to roughly illustrate the theory with an example rather than providing the exact results. We refer to the nice book by Sendra et al. (2007) and the references therein for a detailed view of this theory, from a computer algebra approach.

We focus our attention on curves defined by $f(x, y) = 0$, where f is a polynomial in two variables, and on **affine rational parametrizations**, i.e., parametrizations whose components are rational functions.

Let K be an algebraically closed field of characteristic 0. For example, we might think of $K = \mathbb{C}$.

Definition 1.6.1 (Sendra et al. (2007), Definition 4.1 and Definition 4.12) The affine curve C in the affine plane over K defined by the square-free polynomial $f(x, y)$ is rational (or parametrizable) if there are rational functions $\chi_1(t)$, $\chi_2(t)$ (quotients of polynomials with coefficients in K) such that

- for almost all $t_0 \in K$ (i.e., for all but a finite number of exceptions), the point $(\chi_1(t_0), \chi_2(t_0))$ is on C, and
- for almost every point $(x_0, y_0) \in C$ there is $t_0 \in K$ such that $(x_0, y_0) = (\chi_1(t_0), \chi_2(t_0))$.

Then $(\chi_1(t), \chi_2(t))$ is called an affine rational parametrization of C. $(\chi_1(t), \chi_2(t))$ is in reduced form if the rational functions $\chi_1(t) = \frac{\chi_{11}(t)}{\chi_{12}(t)}$, and $\chi_2(t) = \frac{\chi_{21}(t)}{\chi_{22}(t)}$ are in reduced form; i.e., $\gcd(\chi_{j1}(t), \chi_{j2}(t))$ is trivial for $j = 1, 2$.

An affine parametrization of a rational curve is said to be proper if it admits a rational inverse, or equivalently, if almost every point in the curve is generated by exactly one value of the parameter.

Theorem 1.6.2 (Sendra et al. (2007), Theorem 4.39) *Let $(\chi_1(t), \chi_2(t))$ be a proper parametrization in reduced form of a rational affine plane curve. We write $\chi_j(t) = \frac{\chi_{j1}(t)}{\chi_{j2}(t)}$ for $j = 1, 2$. Then, the implicit polynomial defining the curve is given by*

$$\mathrm{res}_t \left(x\chi_{12}(t) - \chi_{11}(t), \, y\chi_{22}(t) - \chi_{21}(t) \right).$$

The symbol $\text{res}(f, g)$ stands for the **resultant** of $f \in K[x]$ and $g \in K[x]$. Let $f(x) = a_n x^n + \ldots + a_1 x + a_0 \neq 0$, and $g(x) = b_m x^m + \ldots + b_1 x + b_0$, with $a_j \in K$ for $0 \leq j \leq n$, $a_n \neq 0$, and $b_j \in K$ for $0 \leq j \leq m$, with $b_m \neq 0$. Then,

$$\text{res}_x(f, g) = \begin{vmatrix} a_n & \cdots & & a_0 & & & \\ & a_n & \cdots & a_0 & & \\ & & \ddots & \ddots & \ddots & \\ & & & a_n & \ddots & \cdots & a_0 \\ b_m & \cdots & & b_0 & & \\ & b_m & & \cdots & b_0 & \\ & & \ddots & \ddots & \ddots & \\ & & & b_m & \cdots & b_0 \end{vmatrix}.$$

(observe the matrix involved in the resultant has $n + m$ rows and columns, therefore, it belongs to $\mathcal{M}_{(n+m) \times (n+m)}(K)$). The case of real curves is detailed in (Sendra et al. 2007, Chapter 7).

An Example of Application in Architecture

An architect might be interested in incorporating structures satisfying certain physical, aesthetic, etc. properties. In case such properties can be explained with a parametrization, it is of interest to have the knowledge of an implicit equation defining the curve. In the framework of rational algebraic curves, the previous theory is applicable, leading to a non-local solution. The knowledge of the implicit equation, as mentioned above, is useful for checking whether a point belongs to the curve or not, or for determining global topological properties of the curve.

More precisely, the nature of a cycloid is of interest in the design of the vaults of Kimbell Art Museum (see Sect. 1.2). As mentioned above, a cycloid is drawn by the roulette of a fixed point in a circle rolling on a line.

An architect might be interested in the shape of a curve describing the roulette of a fixed point in a parabola rolling around another parabola. The **cissoid of Diocles** is indeed defined in this way: let $y = ax^2$ be a parabola, for some $a > 0$, and consider the parabola $y = -ax^2$. Rolling the first parabola along the second one is equivalent to choosing a moving point in the second parabola and computing the symmetric of the second parabola with respect to the tangent line at such point. The roulette of the vertex in the symmetric parabola draws the curve under inspection (the situation is illustrated in Fig. 1.26). The following link describes the roulette of the vertex of the first parabola as it rolls around the second one (see Fig. 1.27 and the QR code in Fig. 1.25).

Let $P = (t, -at^2)$ be a generic point in the fixed parabola C_1. The tangent line of C_1 at P has equation

$$y + at^2 = (-2at)(x - t).$$

Fig. 1.25 QR Code 4

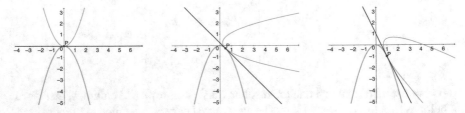

Fig. 1.26 Graph of $y = x^2$ rolling around $y = -x^2$ (symmetric parabola of $y = -x^2$ with respect to the tangent lines)

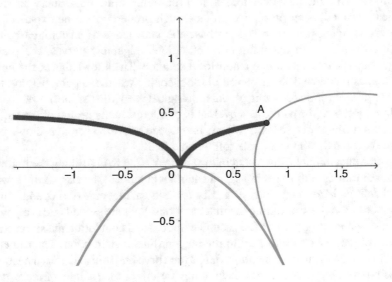

Fig. 1.27 Roulette of a point drawing a cissoid

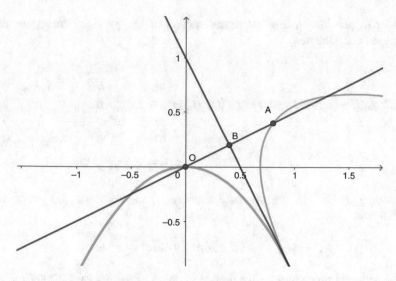

Fig. 1.28 Construction of the cissoid

Direct computations yield the new coordinates of the vertex of the parabola at any time $t \in \mathbb{R}$ (see Fig. 1.28):

$$B = B(t) = \left(\frac{2a^2t^3}{4a^2t^2 + 1}, \frac{at^2}{4a^2t^2 + 1} \right), \quad A = A(t) = \left(\frac{4a^2t^3}{4a^2t^2 + 1}, \frac{2at^2}{4a^2t^2 + 1} \right).$$

This means that a parametrization of the cissoid is

$$\alpha(t) = (x(t), y(t)) = \left(\frac{4a^2t^3}{4a^2t^2 + 1}, \frac{2at^2}{4a^2t^2 + 1} \right), \quad t \in \mathbb{R}.$$

It is not difficult to see that $(\mathbb{R}, \alpha)$ is a proper parametrization. Let us write

$$\chi_1(t) = \frac{\chi_{11}(t)}{\chi_{12}(t)} = \frac{4a^2t^3}{4a^2t^2 + 1}, \quad \chi_2(t) = \frac{\chi_{21}(t)}{\chi_{22}(t)} = \frac{2at^2}{4a^2t^2 + 1}.$$

It holds that

$$\frac{y}{x} = \frac{\chi_2(t)}{\chi_1(t)} = 2at.$$

In order to provide an implicit representation of the previous curve, we apply Theorem 1.6.2. One has

$$\text{res}_t(x\chi_{12}(t) - \chi_{11}(t), y\chi_{22}(t) - \chi_{21}(t)) = \begin{vmatrix} -4a^2 & 4a^2x & 0 & x & 0 \\ 0 & -4a^2 & 4a^2x & 0 & x \\ 4a^2y - 2a & 0 & y & 0 & 0 \\ 0 & 4a^2y - 2a & 0 & y & 0 \\ 0 & 0 & 4a^2y - 2a & 0 & y \end{vmatrix}$$

$$= 16a^4x^2y + 16a^4y^3 - 8a^3x^2.$$

Dividing by $8a^3$ in the previous expression, we conclude the curve under inspection is

$$C = \{(x, y) \in \mathbb{R}^2 : 2ax^2y + 2ay^3 - x^2 = 0\}.$$

The converse procedure is also treated in detail in Sendra et al. (2007): starting from a polynomial in two variables $f(x, y)$, determine whether the set of zeroes of such polynomial describes a rational plane curve, and in case the answer is positive, determine a global rational parametrization of the rational curve.

Definition 1.6.3 (Sendra et al. (2007), Definition 4.15 and Definition 4.16) Let $\chi(t)$ be a rational function in reduced form. If $\chi(t)$ is not zero, the degree of $\chi(t)$ is the maximum of the degrees of the numerator and denominator of $\chi(t)$.

We define the degree of an affine rational parametrization as the maximum of the degrees of its components.

Theorem 1.6.4 (Sendra et al. (2007), Theorem 4.21) *Let C be an affine rational curve defined over K with defining polynomial $f(x, y) \in K[x, y]$, and let $\mathcal{P}(t) = (\chi_1(t), \chi_2(t))$ be a parametrization of C. Then, the parametrization is proper if and only if*

$$deg(\mathcal{P}(t)) = max\{deg_x(f), deg_y(f)\}.$$

Furthermore, if $\mathcal{P}(t)$ is proper and $\chi_1(t)$ is nonzero, then $deg(\chi_1(t)) = deg_y(f)$; similarly, if $\chi_2(t)$ is nonzero then $deg(\chi_2(t)) = deg_x(f)$.

A first example of an application of the previous theory is the following. Assume a curve is given as the locus under certain property. Ellipses, hyperbolas and parabolas can be defined in this way (see Sect. 1.4). Let C be the parabola given by the points of $\mathbb{R}^2$ such that the distance to the line $x + y = 0$ equals the distance to the point $P = (2, 2)$. It holds that

$$C = \{(x, y) \in \mathbb{R}^2 : (x + y)^2 - 2(x - 2)^2 - 2(y - 2)^2 = 0\}.$$

By Theorem 1.6.4, a proper rational parametrization of C has as first and second components a rational function whose numerator and denominator are polynmials of degree 2. It is not difficult to verify that

$$\alpha(t) = \left(\frac{t^2}{16} + \frac{t}{2} + 1, \frac{t^2}{16} - \frac{t}{2} + 1 \right), \quad t \in \mathbb{R}$$

is a rational parametrization of C.

The previous example could have been solved by means of the changes of variables described in Sect. 1.4 at the time of providing a classification of conics.

The conchoid of Nicomedes is a classic plane curve which has been used in 3D anamorphic art, i.e., representing perturbations of the curve giving rise to a 3D-like image. It is the locus of points P obtained by shifting the points of a fixed line proportionally to their distance to the origin of coordinates in some direction. One such curve is defined by the equation $f(x, y) = (x - 1)^2(x^2 + y^2) - x^2$, and a rational parametrization of the curve is given by

$$\alpha(t) = \left(1 + \frac{1 - t^2}{1 + t^2}, 2t \left(\frac{1}{1 - t^2} + \frac{1}{1 + t^2} \right) \right), \quad t \in \mathbb{R}.$$

Figure 1.29 illustrates this particular conchoid of Nicomedes (Figs. 1.30 and 1.31).

Let us study a more complicated example. We focus our attention on certain curves drawn by a fixed point in a circle which rolls inside or outside another

Fig. 1.29 Conchoid of Nicomedes

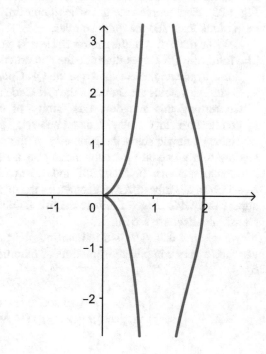

Fig. 1.30 QR Code 5 (left) Deltoid with $R/r = 3$; QR Code 6 (right) Astroid with $R/r = 4$

Fig. 1.31 QR Code 7 (left) Cardioid with $R/r = 1$; QR Code 8 (center) Nephroid with $R/r = 2$; QR Code 9 (right) Epicycloid with $R/r = 3$

circle, which is fixed: **Hypocycloids**, if the rolling circle is inside the fixed one; **epicycloids**, in the case that the rolling circle is outside. Examples of epicycloids are the cardioid (see (1.2) and Fig. 1.2 (left)), nephroid (see Fig. 1.32), and the curve in Fig. 1.2 (right). Examples of hypocycloids are the deltoid and the astroid (see Fig. 1.33). Each curve emerges when departing from different ratios between the radii of the fixed and the rolling circles.

Let r be the radius of the rolling circle and let R be the radius of the fixed circle. The following QR Codes illustrate the construction of some curves of this nature.

Some hypocycloids are shown in the QR Code of Fig. 1.30.

Some epicycloids are shown in the QR Code of Fig. 1.31.

Concerning this situation, this family of curves has been recursively used in architecture. In Čučaković and Paunović (2015), Aleksandar Čučaković and Marijana Paunović study the geometry of the nephroid and the catacaustics grid, leading to a procedure for constructing an anamorphosis of a cube by means of horizontal sections. In Marchetti and Costa (2015) Elena Marchetti and Luisa Rossi Costa describe different geometries in Milan cathedral. These geometries also appear in a wide range of artistic manifestations, for example in rose windows of Gothic architecture (see Fig. 1.34).

We assume that R/r is a rational number; otherwise, the construction would describe a very complicated structure. A parametrization of an epicycloid is given by

$$\begin{cases} x(t) = (R+r)\cos(t) - r\cos\left((1 + \frac{R}{r})t\right) \\ y(t) = (R+r)\sin(t) - r\sin\left((1 + \frac{R}{r})t\right) \end{cases} \tag{1.20}$$

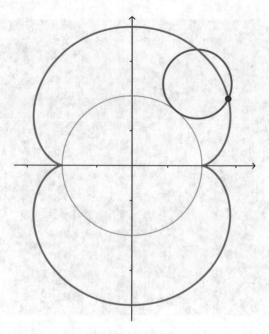

Fig. 1.32 A nephroid as an epicycloid

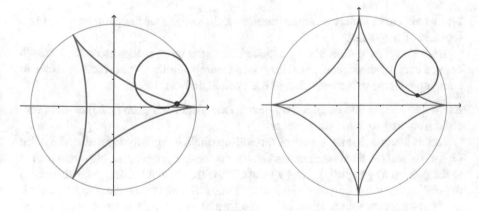

Fig. 1.33 Deltoid (left) and astroid (right) as hypocycloids

The whole curve is drawn for any choice of the domain of the parameter being a segment of length 2π.

A parametrization of an hypocycloid is given by

$$
\begin{cases}
x(t) = (R - r)\cos(t) + r\cos\left(\left(\frac{R}{r} - 1\right)t\right) \\
y(t) = (R - r)\sin(t) - r\sin\left(\left(\frac{R}{r} - 1\right)t\right)
\end{cases}.
\tag{1.21}
$$

Fig. 1.34 Stained glass window from Saint-Chapelle in Paris

The whole curve is drawn for any choice of the domain of the parameter that is a segment of length 2π.

Both parametrizations can be deduced from the geometric scheme. In order to deduce each of them, one can follow a reasoning similar to that used in order to come up with the parametrization of the cycloid, in Sect. 1.2.

Example 1.6.5 In this example, we deduce the parametrization (1.20) of an epicycloid from its geometric construction.

Let us assume that the center of the rolling circle is initially located at the point $(R+r, 0)$, and the fixed point in it is $(R, 0)$. As the center of the rolling circle moves to the point $((R+r)\cos(t), (R+r)\sin(t))$, the fixed point has moved relatively in the rolling circle tR units anticlockwise. These tR units correspond to an angle θ in the rolling circle with $\theta r = tR$. Therefore, $\theta = tR/r$. The fixed point is now located at the point $(-r\cos(\theta + t), -r\sin(\theta + t))$ with respect to the center of the rolling circle. The parametrization of the epicycloid (1.20) is deduced from here.

Example 1.6.6 The parametrization (1.21) of an hypocycloid can be obtained in an analogous manner. Regarding the hypocycloid, the center of the rolling circle is initially at the point $(R - r, 0)$ which moves to $((R - r)\cos(t), (R - r)\sin(t))$. An analogous reasoning as before leads to (1.21).

At this point, we focus on the curve defining a cardioid, for simplicity. It is straightforward to verify that for the case $R = r$, the parametrization in (1.20) reads as follows:

$$\begin{cases} x(t) = 2r\sin(t) - r\sin(2t) \\ y(t) = 2r\cos(t) - r\cos(2t). \end{cases}$$

This is a satisfactory parametrization. However, one may be interested in a rational parametrization of the cardioid. It is straight to check that for $r = 1$, the cardioid is also defined implicitly by

$$C = \{(x, y) \in \mathbb{R}^2 : (x^2 + y^2)^2 + 8y - 6(x^2 + y^2) - 3 = 0\}.$$

Let $f(x, y) = (x^2 + y^2)^2 + 8y - 6(x^2 + y^2) - 3$. By Theorem 1.6.4 we get that any proper parametrization of the cardioid must have first and second components of degree 4. Searching for such polynomials has an enormous computational cost. However, one can follow one of the approaches in Lastra et al. (2018) to obtain that

$$\begin{cases} x(t) = \frac{-2t^4 + 4t^3 - 4t + 2}{(t^2 + 1)^2} \\ y(t) = \frac{t^4 + 4t^3 - 6t^2 + 4t + 1}{(t^2 + 1)^2}, \quad t \in \mathbb{R} \end{cases}$$

is a rational parametrization of the cycloid.

Other geometric objects of a different nature, such as fractals, have been used to inspire the design of rosettes (see Gailiunas 2014), and other architectural elements.

1.7 Approximation and Interpolation of Curves

In practice, the curve fitting our needs does not come from a level curve of a regular function $f(x, y)$. Instead, one **approximates** a given curve by others which are easier to handle, or one searches for a curve passing through certain distinguished points, under other additional properties, via **interpolation**.

A satisfactory answer to the approximation problem is attained with Bézier curves and the algorithm of Boor–de Casteljau. Generally speaking, a curve is approximated constructing a cubic, i.e., a parametrization whose components turn out to be polynomials of degree three in the parameter, starting from four given approximation points. Given the points $A, B, C, D \in \mathbb{R}^2$, the curve constructed by the Boor–de Casteljau algorithm is determined by the parametrization

$$\alpha(s) = (1 - t)^3 A + 3(1 - t)^2 t B + 3(1 - t)t^2 C + t^3 D, \quad t \in (0, 1). \tag{1.22}$$

The algorithm is as follows. Given these four points, one considers for each $s \in (0, 1)$ the point $P_{AB} := At + B(1 - t)$ which belongs to the segment joining the

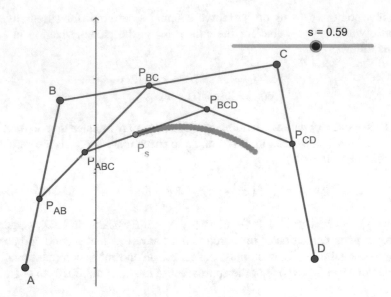

Fig. 1.35 Boor–de Casteljau algorithm

Fig. 1.36 QR Code 10

points A and B. The same construction for the pairs of points B and C, and for C and D is followed to obtain P_{BC} and P_{CD}. We proceed in the same way with the pairs of points P_{AB} and P_{BC}; and P_{BC} together with P_{CD}, leading to the points P_{ABC} and P_{BCD} respectively. We get the final point of the curve $P = P(s)$ by the same argument on the pair of points P_{ABC} and P_{BCD}. Varying the value of the parameter t in $(0, 1)$ we get (1.22). That construction is illustrated in Fig. 1.35.

The QR Code in Fig. 1.36 is an interactive link to the algorithm in Geogebra, which makes it possible to vary the points A, B, C, D.

Observe that the limit points of the curve coincide with the first and last points considered for the approximation.

A useful tool is that of B-splines, which consist of the concatenation of Bézier curves which coincide at the limit points, and such that the concatenation is made under certain regularity assumptions (continuity, derivability, etc.) of the resulting curve. This is, in some sense, an interpolation tool, due to the fact that the resulting curve passes through some prefixed points. A different approximation is that of the least-squares method, which is used increasingly due to the scanners and the vast amount of information obtained by them. Given several points in the plane, the standard **least squares method** provides a line such that the distance from the points given to the line (the regression line) is minimum, i.e., the error of this approximation is minimum. Generalizations can be arranged by substituting a line by a polynomial of larger degree, or even with functions of different nature, such as exponential or logarithmic functions.

A deeper insight on Bézier curves, B-splines and NURBS is detailed in (Pottmann et al. 2007, Chapter 8). The use of robots and computer techniques in architecture and urbanism allows complex forms which give rise to novel geometries, as it is described in Picon (2010). Approximation techniques make it possible to design different shapes of buildings where people will assemble. Examples of this are the Kunsthaus in Graz, by Peter Cook and Colin Fournier. A practical approach to a specific problem appears in Brander et al. (2016), one of a number of examples in this Proceedings volume and the related series.

Observe that B-splines are cubic curves which join at the endpoints with certain regularity. More general configurations can be proposed in this direction, i.e., polynomial parametrizations connected together at the endpoints which admit a certain regularity at these points. Another type of interpolation is that which is provided by a unique polynomial passing through some fixed points. That polynomial has a maximum degree of n when providing $n + 1$ points, but regularity is global in the whole curve. This is known as Lagrange interpolation.

Lagrange Interpolating Polynomial
In Sect. 1.1, we have verified that any regular curve

$$C = \{(x, y) \in \mathbb{R}^2 : f(x, y) = 0\}$$

can be written locally as a function of one of the variables, say $y = g(x)$, depending on the component of ∇f which does not vanish near a point of the curve.

Let $\mathcal{P} = \{(x_j, y_j) : 1 \leq j \leq N\}$ be a set of points in the plane. We assume that $x_i \neq x_j$ for all $i \neq j$. We aim to find a polynomial $y = g(x)$, of minimum degree, such that $g(x_j) = y_j$ for all $1 \leq j \leq N$. In other words, we search for the interpolating polynomial of g at the points in $\mathcal{P}$.

The resulting polynomial is known as a Lagrange interpolating polynomial, and it is given by

$$p(x) = \sum_{j=1}^{N} y_j \frac{(x - x_1)(x - x_2) \cdots (x - x_{j-1})(x - x_{j+1}) \cdots (x - x_N)}{(x_j - x_1)(x_j - x_2) \cdots (x_j - x_{j-1})(x_j - x_{j+1}) \cdots (x_j - x_N)}.$$

(1.23)

Theorem 1.7.1 *A Lagrange interpolating polynomial* $p = p(x)$ *is the only polynomial of degree at most* $N - 1$, *such that*

$$g(x_j) = y_j, \quad 1 \leq j \leq N.$$

(1.24)

Proof Observe that for all $1 \leq j \leq N$, the expression

$$g_j(x) = \frac{(x - x_1)(x - x_2) \cdots (x - x_{j-1})(x - x_{j+1}) \cdots (x - x_N)}{(x_j - x_1)(x_j - x_2) \cdots (x_j - x_{j-1})(x_j - x_{j+1}) \cdots (x_j - x_N)}$$

defines a polynomial of degree $N - 1$ which satisfies the condition that $g_j(x_j) = 1$ and $g_j(x_k) = 0$ for all $1 \leq k \leq N$ with $k \neq j$. From the definition of a Lagrange interpolating polynomial in (1.23), we derive (1.24). Let $h(x) = a_0 + a_1 x + \ldots + a_{N-1} x^{N-1}$ be any other polynomial of degree $N - 1$ satisfying (1.24). Then, the vector of coefficients $x = (a_0, a_1, \cdots, a_{N-1})$ is the solution of the linear system $Ax^T = b$, where

$$A = \begin{pmatrix} 1 & x_1 & x_1^2 & \cdots & x_1^{N-1} \\ 1 & x_2 & x_2^2 & \cdots & x_2^{N-1} \\ \vdots & & \ddots & & \vdots \\ 1 & x_N & x_N^2 & \cdots & x_N^{N-1} \end{pmatrix}. \quad b = \begin{pmatrix} y_1 \\ y_2 \\ \vdots \\ y_N \end{pmatrix}.$$

The matrix A is invertible because it is a Van der Monde matrix, and $x_i \neq x_j$ for $i \neq j$. There is a unique solution of the system determined by $x^T = A^{-1}b$, which must coincide with the coefficients of Lagrange interpolating polynomial.

This theory can be applied in architecture in the case where an interesting curve cannot be described exactly by an implicit equation or by an exact parametrization. Let us consider the Teatro Popular in Niterói, Brazil, designed by Oscar Niemeyer (see Fig. 1.37).

Fig. 1.37   Teatro popular in Niterói, Brazil, by Oscar Niemeyer

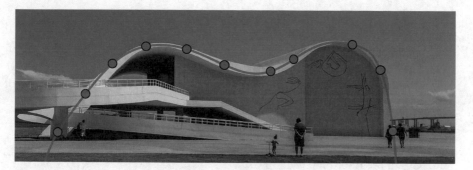

Fig. 1.38 The Teatro popular with author's overlay

The approximation of one of the curves defining this building can be done by means of Lagrange polynomials. In Fig. 1.38, we have considered different points in the curve and approximated it by a polynomial of degree 10. The software for computating the interpolating polynomial is Geogebra, which incorporates a tool that provides the interpolatory polynomial of optimal degree, given some points in the curve.

In practice, other more practical structures such as the algorithm of divided differences are usually applied. Other interpolating families might be appropriate depending on the curve being approximated. For example, the family of functions $\{a_k \cos(b_k x), c_k \sin(d_k x)\}_{k \in \mathbb{N}}$ is also of great interest.

Recent studies use approximation of curves in architecture (Shen et al. 2021) for the parametrization of the curvilinear roofs of traditional chinese architecture.

1.8 Suggested Exercises

1.1. Consider the locus of the point P such that the sum of the square of distances to the points $P_1 = (-\sqrt{2}, 0)$ and $P_2 = (\sqrt{2}, 0)$ is equal to 6. Determine the curve defined by the previous description. Which geometric form does it have? Determine the locus when substituting the word "sum" by "difference" in the previous statement.

1.2. Let $f : \mathbb{R}^2 \to \mathbb{R}$ be the function $f(x, y) = \exp(x - y)$ for all $(x, y) \in \mathbb{R}^2$. Verify that $f \in C^\infty(\mathbb{R}^2)$ and compute its derivatives of any order.

1.3. Verify that any polynomial $p : \mathbb{R}^2 \to \mathbb{R}$ is such that $p \in C^\infty(\mathbb{R}^2)$.

1.4. Let $I \subseteq \mathbb{R}$ be an open interval, and let $g : I \to \mathbb{R}$ such that $g \in C^\infty(I)$. The **graph** of g consists of the set

$$C = \{(x, y) \in I \times \mathbb{R} : y - g(x) = 0\}.$$

Verify that the graph of g is a regular curve and determine a unique parametrization describing such curve.

1.5. Consider the parametrization $(\mathbb{R}, \alpha)$ with $\alpha(t) = (t^2, t^3)$, for all $t \in \mathbb{R}$. Verify that the parametrized curve is regular except from the point $(0, 0)$. Sketch the graph of the curve and observe the form of the curve near that point.

1.6. Find the tangent and normal lines at the regular points of the lemniscate defined in implicit form (see (1.1)) and also in parametric form (see (1.3) and (1.4)).

1.7. Complete the details on the implict expression defining the lemniscate of Bernoulli. Verify that the parametric expression in (1.8) corresponds to that curve.

1.8. Give a proof for Proposition 1.3.10.

1.9. Determine the arc length of a cycloid between the values of the parameter $t = 0$ and $t = 2\pi$.

1.10. Determine the arc length of a logarithmic spiral, parametrized in (1.9), between any pair of points.

1.11. Determine the curvature of the curve defined by the graph of a $\mathcal{C}^\infty(I)$ function at each of its points.

1.12. Determine the curvature at every point of the ellipse defined by

$$C = \{(x, y) \in \mathbb{R}^2 : \frac{x^2}{a^2} + \frac{y^2}{b^2} - 1 = 0\},$$

for every fixed $a, b > 0$.

1.13. Do the same for the hyperbola

$$C = \{(x, y) \in \mathbb{R}^2 : \frac{x^2}{a^2} - \frac{y^2}{b^2} - 1 = 0\},$$

for every fixed $a, b > 0$.

1.14. Verify that the conic

$$C = \{(x, y, z) \in \mathbb{R}^3 : x^2 + y^2 + 2xy + 2x + 1 = 0\}$$

is a parabola. Find its vertex from its line of symmetry.

1.15. Give details about the procedure to obtain the classification of the reducible conics.

1.16. Determine the factorization of the conic C_3 in Example 1.4.5 as suggested in that example. Take into account the degrees of freedom that can come up in the process.

1.17. Consider the conic of equation

$$C = \{(x, y) \in \mathbb{R}^2 : x^2 + y^2 + 4xy + 4x - 2y + 1 = 0\}.$$

Classify the conic and determine its characteristic elements.

1.18. Consider the conic of equation

$$C = \{(x, y) \in \mathbb{R}^2 : 2x^2 + 2y^2 + 2xy + 2x + 2y - 1 = 0\}.$$

Classify the conic and determine its characteristic elements.

1.19. Verify that

$$(x(\theta), y(\theta)) = \left(\frac{\cos(\theta)}{\sqrt{6}} + \frac{\sin(\theta)}{\sqrt{2}}, \frac{\cos(\theta)}{\sqrt{6}} - \frac{\sin(\theta)}{\sqrt{2}} \right), \quad \theta \in \mathbb{R} \qquad (1.25)$$

parametrizes the conic

$$C = \{(x, y) \in \mathbb{R}^2 : 2x^2 + 2y^2 + 2xy - 1 = 0\}.$$

Follow the algorithm of parametrization of a conic stated in Sect. 1.4 in order to obtain such parametrization. Classify the conic and determine its characteristic elements. Obtain the normal and tangent line associated to the curve at the point $P = (1/\sqrt{6}, 1/\sqrt{6})$ from the parametrization, and also from the implicit equation determining the conic.

1.20. Use the techniques of implicitation of a rational parametrization in Sect. 1.6 to derive the implicit equation of the lemniscate, given in (1.8). Hint: A parametrization of the unit circle is $(x(s), y(s)) = ((1-s^2)/(1+s^2), 2s/(1+s^2))$. Hint: Make use of a symbolic computation program such as Calcme: https://calcme.com

1.21. Let us consider the unit circle $C = \{(x, y) \in \mathbb{R}^2 : x^2+y^2-1 = 0\}$. Determine a parametrization of the unit circle by considering the parametrization of a pencil of lines through the point $(1, 0)$ and computing the intersection with C. Compare with the hint given in the previous exercise for the implicitation of the parametrization of a lemniscate.

1.22. Id. for the ellipse $C_2 = \{(x, y) \in \mathbb{R}^2 : \frac{x^2}{a^2} + \frac{y^2}{b^2} - 1 = 0\}$ and the point $(a, 0)$.

1.23. Id. for the hyperbola $C_2 = \{(x, y) \in \mathbb{R}^2 : \frac{x^2}{a^2} - \frac{y^2}{b^2} - 1 = 0\}$ and the point $(a, 0)$.[2]

1.24. Compute the Lagrange interpolating polynomial of degrees $N = 0, 1, 2$ approximating the catenary $y = \frac{a}{2}(\exp(\frac{x}{a}) + \exp(-\frac{x}{a}))$ at equidistant points in the interval $[0, 1]$.

---

[2]A more general procedure to obtain the parametrization by lines of a given irreducible conic and other classes of curves can be found in Section 4.6 of Sendra et al. (2007).

Chapter 2
Parametrizations and Space Curves

This chapter is devoted to the study of space curves and their application to architectural elements. It can be viewed as an extension of the theory stated in Chap. 1, considering not only curves which lie inside a plane but also those in Euclidean space. The reader will find a parallelism between the arrangement of the concepts in Chap. 1 and this one.

Several concepts in plane differential geometry of curves extend themselves naturally to the study of space curves, such as the tangent line. However, others arise from this more complicated and rich topic. As we have already mentioned, the definition of a curve makes it possible to adopt aesthetic or physical properties leading to certain luminosity of the space, load supports, etc.

We detail the concepts of space curve and parametrization of a curve. Moreover, we analyse the coordinate system leading to Frenet-Serret formulas, associated to regular curves, and concepts such as curvature and torsion. We also focus on some application of these concepts and useful mathematical techniques which can be applied in architecture. Here we put stress on helices and related objects, which are of frequent use in architecture and art.

2.1 Space Curves and Parametrizations

The curves under consideration in the previous chapter had in common their planar nature, and where defined as the image of a parametrization or in implicit form. Now, space curves represent the trajectory of a moving point in the three-dimensional space.

A circle at certain fixed height (see Fig. 2.1, left) is a space curve. However, it can be considered as a plane curve, restricting its study to the plane containing the curve. However, a helix (see Fig. 2.1, right) is a space curve which can not be embedded in a plane.

© The Author(s), under exclusive license to Springer Nature Switzerland AG 2021
A. Lastra, *Parametric Geometry of Curves and Surfaces*, Mathematics and the Built Environment 5, https://doi.org/10.1007/978-3-030-81317-8_2

Fig. 2.1 Circle centered at $P = (0, 0, 1)$ and radius $R = 3$, at height $z = 1$ (left); Helix (right)

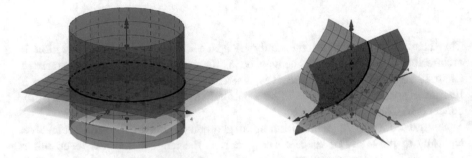

Fig. 2.2 Curves in Fig. 2.1 determined by the intersection of surfaces

In the previous examples, the circle was defined as the set of points

$$C_1 = \{(x, y, z) \in \mathbb{R}^3 : x^2 + y^2 - 9 = 0, z - 1 = 0\},$$

whilst the helix is determined by

$$C_2 = \{(x, y, z) \in \mathbb{R}^3 : x - \cos(z) = 0, y - \sin(z) = 0\}.$$

This approach indicates one of the ways that can be followed to define a space curve. In Chap. 3, it will become clear that the curves defined in this manner are determined by the intersection of two surfaces (see Fig. 2.2).

Definition 2.1.1 Let $\emptyset \neq U \subseteq \mathbb{R}^3$ be an open set and let $f_1, f_2 : U \to \mathbb{R}$ be two functions with $f_1, f_2 \in C^\infty(U)$. Assume that the set

$$C = \{(x, y, z) \in U : f_1(x, y, z) = 0, f_2(x, y, z) = 0\}$$

is not empty. Then we say that C is a **space curve**.

Fig. 2.3 Viviani's curve (left), and solenoid toric (right)

A different example is **Viviani's curve**. This curve is defined by the $(x, y, z) \in \mathbb{R}^3$ such that

$$(x, y, z) = (1 + \cos(t), \sin(t), 2 \sin\left(\frac{t}{2}\right)), \quad t \in \mathbb{R}, \tag{2.1}$$

(see Fig. 2.3, left). A **loxodrome** (or **rhumb line**) is a curve which always follows the same direction. One of these curves is the **solenoid toric** (see Fig. 2.3, right) which is determined by the points $(x, y, z) \in \mathbb{R}^3$ satisfying

$$(x, y, z) = ((R + r\cos(nt))\cos(t), (R + r\cos(nt))\sin(t), r\sin(nt)), \quad t \in \mathbb{R}, \tag{2.2}$$

for some fixed $R > r > 0$ and $n \in \mathbb{N}$.

Definition 2.1.2 A **parametrization** is a pair (I, α), where $I \subseteq \mathbb{R}$ is an open interval and $\alpha : I \to \mathbb{R}^3$ belongs to $\mathcal{C}^\infty(I)$.

Again, the regularity conditions stated in Definitions 2.1.1 and 2.1.2 can be weakened in the results presented, but we have maintained them for the sake of simplicity. The concept of a regular parametrization can be naturally extended from the framework of planar curves (see Definition 1.1.3). The first condition in the next definition is linked to the possibility of defining a tangent line at every point of the curve whilst the second condition there excludes the appearance of autointersections.

Definition 2.1.3 The parametrization (I, α) is a **regular parametrization** if it satisfies the conditions that

- $\alpha'(t) \neq (0, 0, 0)$ for all $t \in I$,
- $\alpha : I \to \mathbb{R}^3$ is a one-to-one function.

Both conditions are blended with a three-dimensional version of Definition 1.1.5 in order to provide the notion of a regular space curve. In this regard, the condition of the existence of a disc in Definition 1.1.5 assures that the curve is essentially a "bent" segment, when regarded it locally.

Let us consider the parametrization (I, α) defined by $\alpha(t) = (t^2, t^3, t^4)$, for $t \in I = (-2, 2)$. We have $\alpha'(0) = (0, 0, 0)$. The splitting of (I_1, α_1) and (I_2, α_2)

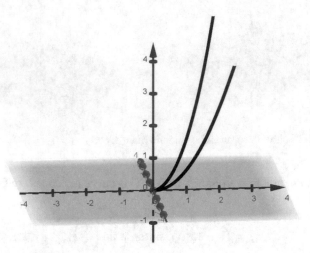

Fig. 2.4 $\alpha(t) = (t^2, t^3, t^4), t \in (-2, 2)$

determined by $\alpha_1(t) := \alpha(t)$ for $t \in I_1 = (-2, 0)$, and $\alpha_2(t) := \alpha(t)$ for $t \in I_2 = (0, 2)$ parametrizes $\alpha(I)$ except for the point $(0, 0, 0) = \alpha(0)$. Observe in Fig. 2.4 the behavior of the curve at $P = (0, 0, 0)$.

In the case of Viviani's curve (see Fig. 2.3 (left) and (2.1)), the point $P = (2, 0, 0)$ presents an autointersection.

Definition 2.1.4 Given a curve C, we say that the parametrization (I, α) is a parametrization of the curve C if $C = \alpha(I)$.

Given a parametrization (I, α), the set $\alpha(I) \subseteq \mathbb{R}^3$ is said to be an **arc**, which is a **regular arc** if the parametrization associated to it is regular.

The following example explains the definition of regular space curve. Also, analogous examples as Example 1.1.10 can be found in this framework.

Example 2.1.5 The curve

$$C = \{(x, y, z) \in \mathbb{R}^3 : (x^2 + y^2 + 2x)^2 - 4(x^2 + y^2) = 0; 4z - x^2 - y^2 = 0\}$$

can be parametrized by

$$\alpha(t) = (2(1 - \cos(t)) \cos(t), 2(1 - \cos(t)) \sin(t), (1 - \cos(t))^2), \quad t \in (0, 2\pi)$$

which draws the whole curve. Figure 2.5 shows the curve from different angles. Note that this curve is not contained in any plane in $\mathbb{R}^3$, and $\alpha'(0) = (0, 0, 0)$, with $\alpha(0) = (0, 0, 0)$.

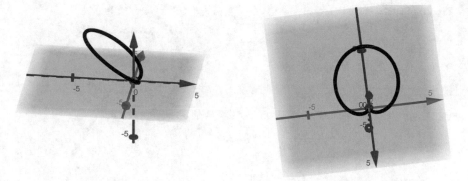

Fig. 2.5 Curve in Example 2.1.5

Fig. 2.6 Parametrizations (I_1, α_1) and (I_2, α_2) associated to Viviani's curve

In the case of Viviani's curve the parametrizations (I_1, α_1) and (I_2, α_2) determined by

$$\alpha_1(t) = (1 + \cos(t), \sin(t), 2\sin(t/2)), \quad t \in I_1 = (0, 2\pi), \tag{2.3}$$

$$\alpha_2(t) = (1 + \cos(t), \sin(t), 2\sin(t/2)), \quad t \in I_2 = (-2\pi, 0), \tag{2.4}$$

describe the whole curve, except from $P = (2, 0, 0)$. See Fig. 2.6.

Definition 2.1.6 A set $\emptyset \neq C \subseteq \mathbb{R}^3$ is a **regular (space) curve** if for every $(x_0, y_0, z_0) \in C$, there exists a ball $D((x_0, y_0, z_0), r) \subseteq \mathbb{R}^3$, such that $D((x_0, y_0, z_0), r) \cap C$ is a regular arc, i.e., there exists a regular parametrization (I, α) with $\alpha(I) = D((x_0, y_0, z_0), r) \cap C$.

Concerning space curves, Theorem 1.1.6 reads as follows, rewritten in the form of Theorem 2.1.8 with the notations adopted (Fig. 2.7).

Theorem 2.1.7 *Let $U \subseteq \mathbb{R}^3$ be a nonempty open set of $\mathbb{R}^3$. Let $f : U \to \mathbb{R}$ with $f \in \mathcal{C}^\infty(U)$, and $P = (x_0, y_0, z_0) \in U$ such that*

$$rank \begin{pmatrix} \frac{\partial f_1}{\partial x}(P) & \frac{\partial f_1}{\partial y}(P) & \frac{\partial f_1}{\partial z}(P) \\ \frac{\partial f_2}{\partial x}(P) & \frac{\partial f_2}{\partial y}(P) & \frac{\partial f_2}{\partial z}(P) \end{pmatrix} = 2.$$

Fig. 2.7 Regular curve

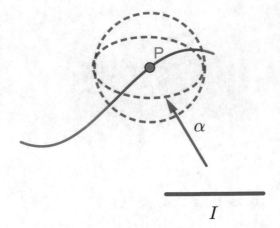

We consider the nonempty set

$$C = \{(x, y, z) \in \mathbb{R}^3 : f_1(x, y, z) - f_1(P) = 0, \ f_2(x, y, z) - f_2(P) = 0\}.$$

Then there exists $r > 0$, an open interval $I \subseteq \mathbb{R}$ and a one-to-one function α : $I \to \mathbb{R}^3$, with $\alpha \in C^\infty(I)$, with $\alpha'(t) \neq (0, 0, 0)$ for all $t \in I$ such that $\alpha(I) = D(P, r) \cap C$.

Theorem 2.1.8 *Let $U \subseteq \mathbb{R}^3$ be a nonempty open set of $\mathbb{R}^3$. Let $f : U \to \mathbb{R}$ with $f \in C^\infty(U)$. We consider the set*

$$C = \{(x, y, z) \in U : f(x, y, z) = 0\}.$$

If $C \neq \emptyset$ and for every $P \in C$ it holds that

$$rank \begin{pmatrix} \frac{\partial f_1}{\partial x}(P) & \frac{\partial f_1}{\partial y}(P) & \frac{\partial f_1}{\partial z}(P) \\ \frac{\partial f_2}{\partial x}(P) & \frac{\partial f_2}{\partial y}(P) & \frac{\partial f_2}{\partial z}(P) \end{pmatrix} = 2, \tag{2.5}$$

then C is a regular space curve.

Outline of the Proof We only give some details of the proof, which is analogous to that of Theorem 1.1.7. Let $P = (x_0, y_0, z_0) \in C$. we can assume, without loss of generality, that

$$\begin{vmatrix} \frac{\partial f_1}{\partial x}(P) & \frac{\partial f_1}{\partial y}(P) \\ \frac{\partial f_2}{\partial x}(P) & \frac{\partial f_2}{\partial y}(P) \end{vmatrix} \neq 0.$$

The continuity of the partial derivatives of f guarantee that the previous determinant is not zero when evaluated at points on some neighborhood of P in U. The implicit mapping theorem guarantees the existence of $I \subseteq \mathbb{R}$ with $z_0 \in I$ and $g_1, g_2 : I \to$

$\mathbb{R}$, g_1, $g_2 \in C^\infty(I)$ and $g_1(z_0) = x_0$, $g_2(z_0) = y_0$ such that $f(g_1(z), g_2(z), z) = 0$ for all $z \in I$. We consider the pair (I, α), with $\alpha(t) = (g_1(t), g_2(t), t)$, for $t \in I$. That pair is a parametrization of an arc contained in the curve. The functions g_1, $g_2 \in C^\infty(I)$ by virtue of the implicit mapping theorem. Moreover, it is clear that $\alpha'(t) \neq (0, 0, 0)$ for all $t \in I$. The existence of a ball $D = D((x_0, y_0, z_0), r) \subseteq \mathbb{R}^3$ with $D \cap C$ being a regular arc is a consequence of the implicit mapping theorem, in which $r > 0$ can be reduced, if necessary.

The reciprocal result is also available, as it was for plane curves.

Theorem 2.1.9 *Let C be a regular curve. For every $(x_0, y_0, z_0) \in C$ there exists $r > 0$, and a vectorial function $f = (f_1, f_2) : D((x_0, y_0, z_0), r) \to \mathbb{R}^2$, $f_1, f_2 \in C^\infty(D((x_0, y_0, z_0), r))$, such that*

$$C \cap D((x_0, y_0, z_0), r) = \{(x, y, z) \in D((x_0, y_0, z_0), r) : f_1(x, y, z) = 0, f_2(x, y, z) = 0\},$$

and

$$rank \begin{pmatrix} \frac{\partial f_1}{\partial x}(Q) & \frac{\partial f_1}{\partial y}(Q) & \frac{\partial f_1}{\partial z}(Q) \\ \frac{\partial f_2}{\partial x}(Q) & \frac{\partial f_2}{\partial y}(Q) & \frac{\partial f_2}{\partial z}(Q) \end{pmatrix} = 2,$$

for all $Q \in D((x_0, y_0, z_0), r)$.

Outline of the Proof The proof follows steps similar to that of Theorem 1.1.8. Given $(x_0, y_0, z_0) \in C$, we consider $D((x_0, y_0, z_0), r_1) \subseteq \mathbb{R}^3$ with $D((x_0, y_0, z_0), r_1) \cap C$ being an arc of regular curve. There exists a regular parametrization (I, α), with $\alpha(I) = D((x_0, y_0, z_0), r_1) \cap C$. Let $\alpha = (\alpha_1, \alpha_2, \alpha_3)$ and assume that $t_0 \in I$ is such that $\alpha(t_0) = (x_0, y_0, z_0)$.

From the continuity of α_1 we obtain that $\alpha_1'(t) \neq 0$ for all $t \in I_1 \subseteq I$, for some open interval I_1, with $t_0 \in I_1$. The function α_1 is invertible in I_1. Let $\alpha_1^{-1} : I_2 \to I_1$, with $x_0 \in I_2 \subseteq \mathbb{R}$ being an open interval of $\mathbb{R}$.

The proof concludes with the choice of the function $f : D((x_0, y_0, z_0), r) \to \mathbb{R}^2$, for some adequate $r_1 \geq r > 0$, given by

$$f(x, y, z) = (y - \alpha_2(\alpha_1^{-1}(x)), z - \alpha_3(\alpha_1^{-1}(x))).$$

Observe that if $Q = (x, y, z) \in D((x_0, y_0, z_0), r)$, then

$$rank \begin{pmatrix} \frac{\partial f_1}{\partial x}(Q) & \frac{\partial f_1}{\partial y}(Q) & \frac{\partial f_1}{\partial z}(Q) \\ \frac{\partial f_2}{\partial x}(Q) & \frac{\partial f_2}{\partial y}(Q) & \frac{\partial f_2}{\partial z}(Q) \end{pmatrix} = rank \begin{pmatrix} -\alpha_2'(\alpha_1^{-1}(x))(\alpha_1^{-1})'(x) & 1 & 0 \\ -\alpha_3'(\alpha_1^{-1}(x))(\alpha_1^{-1})'(x) & 0 & 1 \end{pmatrix} = 2.$$

Theorem 2.1.9 leads to the next definition.

Definition 2.1.10 Let $U \subseteq \mathbb{R}^3$ be a nonempty open set of $\mathbb{R}^3$. Let $f = (f_1, f_2) :$
$U \to \mathbb{R}^2$ with $f_1, f_2 \in C^\infty(U)$. We say that

$$C = \{(x, y, z) \in U : f_1(x, y, z) = 0, \ f_2(x, y, z) = 0\}$$

is a **regular curve** if $C \neq \emptyset$ and for all $P \in C$

$$\text{rank} \begin{pmatrix} \frac{\partial f_1}{\partial x}(P) & \frac{\partial f_1}{\partial y}(P) & \frac{\partial f_1}{\partial z}(P) \\ \frac{\partial f_2}{\partial x}(P) & \frac{\partial f_2}{\partial y}(P) & \frac{\partial f_2}{\partial z}(P) \end{pmatrix} = 2. \tag{2.6}$$

2.2   Some Elements of Regular Space Curves

So far, a parallel theory has been shown regarding plane regular curves in Sects. 1.1
and 2.1. In both of them, regular curves have arisen from implicit or parametric
expressions. In this section, we state generalizations of the elements described in
Sect. 1.3 in the framework of space curves. The enhanced complexity of space
curves with respect to plane curves, makes it necessary to introduce new concepts
such as the torsion of a curve at a point.

Local parametrizations seem to be the answer to separate sensitive points in a
curve. In this respect, we consider the curve studied in Example 2.1.5, and focus on
the point $P = (0, 0, 0)$. Let

$$f_1(x, y, z) = (x^2 + y^2 + 2x)^2 - 4(x^2 + y^2), \quad f_2(x, y, z) = 4z - x^2 - y^2.$$

It holds that

$$\text{rank} \begin{pmatrix} \frac{\partial f_1}{\partial x}(P) & \frac{\partial f_1}{\partial y}(P) & \frac{\partial f_1}{\partial z}(P) \\ \frac{\partial f_2}{\partial x}(P) & \frac{\partial f_2}{\partial y}(P) & \frac{\partial f_2}{\partial z}(P) \end{pmatrix} = \text{rank} \begin{pmatrix} 0 & 0 & 0 \\ 0 & 0 & 4 \end{pmatrix} = 1.$$

From the point of view of parametrizations, one can consider that in Example 2.1.5.
In this case, one has $\alpha(0) = (0, 0, 0) = P$. However,

$$\alpha'(t) = (2\sin(t)(2\cos(t)-1), \ 4\sin^2(t)+2\cos(t)-2, \ 2(1-\cos(t))\sin(t)), \quad t \in \mathbb{R}.$$

This means that $\alpha'(0) = (0, 0, 0)$. The shape of this 3D-cardioid like curve shows a
particular form at P (see Fig. 2.5). The three components of $\alpha'(t)$ vanish at $t = 0$,
and

$$\alpha''(t) = (2\cos(t)(2\cos(t)-1)-4\sin^2(t), \ 8\sin(t)\cos(t)-2\sin(t), \ 4\sin^2(t)), \quad t \in \mathbb{R},$$

which yields $\alpha''(0) \neq (0, 0, 0)$.

Other points which are going to be avoided in our study are the autointersection points. Local parametrizations help in this concern. For example, consider Viviani's curve (see Fig. 2.6), and take the parametrizations (2.3) and (2.4) of arcs contained in the curve. The point $P = (2, 0, 0)$ is not attained by them, but both describe a regular curve, whose image covers almost every point of the curve.

Definition 2.2.1 Given a curve C, a point $P \in C$ is a **singular point** of C if there does not exist $r > 0$ such that $D(P, r) \cap C$ is a regular arc. The points of a curve which are not singular are said to be **regular points** of the curve.

Example 2.2.2 Let C be the set

$$C = \{(x, y, z) \in \mathbb{R}^3 : x^2 - y + z = 0, x^3 - y + z^2 = 0\}.$$

Let $f_1(x, y, z) = x^2 - y + z$ and $f_2(x, y, z) = x^3 - y + z^2$. We search for the singular points $P = (x, y, z) \in \mathbb{R}^3$ of the curve, i.e., the points of the curve such that

$$\mathrm{rank} \begin{pmatrix} \frac{\partial f_1}{\partial x}(P) & \frac{\partial f_1}{\partial y}(P) & \frac{\partial f_1}{\partial z}(P) \\ \frac{\partial f_2}{\partial x}(P) & \frac{\partial f_2}{\partial y}(P) & \frac{\partial f_2}{\partial z}(P) \end{pmatrix} = 2,$$

i.e., the points $(x, y, z) \in \mathbb{R}^3$ such that

$$\mathrm{rank} \begin{pmatrix} 2x & -1 & 1 \\ 3x^2 & -1 & 2z \end{pmatrix} = 2.$$

The determinant of the two last columns in the previous matrix is $-2z + 1$ which vanishes for $z = -1/2$. The other two minors are $-2x + 3x^2$ and $4xz - 3x^2$. In order that the three of them are null, either $x = 0$ or $x = 2/3$ and $z = 1/2$. The values of x, y and z for which the rank of the previous matrix is 1 are either $x = 0$, $y \in \mathbb{R}$ and $z = 1/2$ or $x = 2/3$, $y \in \mathbb{R}$ and $z = 1/2$. Let us consider the points of coordinates $(0, y, 1/2)$ and $(2/3, y, 1/2)$: none of them are points which satisfy both equations of the curve, so they are points which do not belong to the curve. Therefore, the curve does not have singular points.

The definition of velocity vector and tangent line (see Definition 1.3.2) are generalized to regular space curves directly.

Definition 2.2.3 Let (I, α) be a regular parametrization, and $t_0 \in I$, with $P = \alpha(t_0)$. The **velocity vector** (or tangent vector) associated to (I, α) at P is defined by $\alpha'(t_0) \in \mathbb{R}^3$.

The line at $P = \alpha(t_0)$ and direction vector given by its velocity vector $\alpha'(t_0)$ is known as the **tangent line** at P, associated to (I, α).

Proposition 2.2.4 *The tangent line of a regular curve C at a point does not depend on the regular parametrization of the curve.*

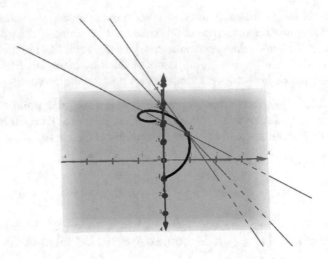

Fig. 2.8 Secant lines

Let $P \in C$, and (I, α) be a regular parametrization of an arc of C with $P \in \alpha(I)$. Then, the tangent line of C at P is given by

$$(x(t), y(t), z(t)) = P + \alpha'(t_0)t, \quad t \in \mathbb{R}$$

with $P = \alpha(t_0)$, for $t_0 \in I$.

The proof of the first part of the previous result is analogous to that of Proposition 1.3.3. In addition, one also obtains a geometric interpretation of the tangent line at a point P as the limit of the secant lines when the secant points tend to P. The construction of the tangent line is adapted from Definition 1.3.5 in this framework (Fig. 2.8).

In the case that the regular curve is determined in implicit form, the tangent line results from two intersecting planes.

Proposition 2.2.5 *Let $C = \{(x, y, z) \in \mathbb{R}^3 : f_1(x, y, z) = 0, f_2(x, y, z) = 0\}$ be a regular curve. For every $P = (x_0, y_0, z_0) \in C$, the tangent line of C at P is given by*

$$\begin{cases} \dfrac{\partial f_1}{\partial x}(P)(x - x_0) + \dfrac{\partial f_1}{\partial y}(P)(y - y_0) + \dfrac{\partial f_1}{\partial z}(P)(z - z_0) = 0 \\ \dfrac{\partial f_2}{\partial x}(P)(x - x_0) + \dfrac{\partial f_2}{\partial y}(P)(y - y_0) + \dfrac{\partial f_2}{\partial z}(P)(z - z_0) = 0. \end{cases}$$

Proof For every $P \in C$ there exists a local regular parametrization of C covering P. Let (I, α) be one of them. It holds that $\alpha(t_0) = P$ for some $t_0 \in I$, and

$f_1(\alpha(t)) = f_2(\alpha(t)) = 0$ for all $t \in I$. Taking derivatives of the previous expression and evaluating at $t = t_0$, we conclude that

$$\left(\frac{\partial f_1}{\partial x}(\alpha(t_0)), \frac{\partial f_1}{\partial y}(\alpha(t_0)), \frac{\partial f_1}{\partial z}(\alpha(t_0)) \right) \cdot \alpha'(t_0) \equiv 0,$$

$$\left(\frac{\partial f_2}{\partial x}(\alpha(t_0)), \frac{\partial f_2}{\partial y}(\alpha(t_0)), \frac{\partial f_3}{\partial z}(\alpha(t_0)) \right) \cdot \alpha'(t_0) \equiv 0. \quad (2.7)$$

Observe from the condition (2.6) that the intersection of the planes of equations $\frac{\partial f_1}{\partial x}(P)(x - x_0) + \frac{\partial f_1}{\partial y}(P)(y - y_0) + \frac{\partial f_1}{\partial z}(P)(z - z_0) = 0$ and $\frac{\partial f_2}{\partial x}(P)(x - x_0) + \frac{\partial f_2}{\partial y}(P)(y - y_0) + \frac{\partial f_2}{\partial z}(P)(z - z_0) = 0$ always determines a line. Such line has $\alpha'(t_0)$ as the directing vector and $\alpha(t_0)$ belongs to it.

Example 2.2.6 Let C be the helix defined by

$$C = \{(x, y, z) \in \mathbb{R}^3 : \cos(z) = x; \sin(z) = y\},$$

parametrized by $(\mathbb{R}, \alpha)$, with $\alpha(t) = (\cos(t), \sin(t), t)$, for $t \in \mathbb{R}$. One can easily verify that the tangent line at any point of the curve coincides either calculating it from its implicit or from its parametric definition.

Let $f_1(x, y, z) = \cos(z) - x$ and $f_2(x, y, z) = \sin(z) - y$. For every $P = (x_0, y_0, z_0) \in C$ we have

$$\begin{pmatrix} \frac{\partial f_1}{\partial x}(P) & \frac{\partial f_1}{\partial y}(P) & \frac{\partial f_1}{\partial z}(P) \\ \frac{\partial f_2}{\partial x}(P) & \frac{\partial f_2}{\partial y}(P) & \frac{\partial f_2}{\partial z}(P) \end{pmatrix} = \begin{pmatrix} -1 & 0 & -\sin(z_0) \\ 0 & -1 & \cos(z_0) \end{pmatrix},$$

which is of rank 2. The tangent line is defined by

$$\begin{cases} -(x - x_0) - \sin(z_0)(z - z_0) = 0 \\ -(y - y_0) + \cos(z_0)(z - z_0) = 0. \end{cases}$$

On the other hand, if $P = (x_0, y_0, z_0) = \alpha(z_0)$, then the tangent line is parametrized by

$$(x, y, z) = (\cos(z_0), \sin(z_0), z_0) + s(-\sin(z_0), \cos(z_0), 1), \quad s \in \mathbb{R}.$$

Solving the system determined by the previous parametrization in s we get the tangent line as it was derived from the implicit form of the curve, taking into account that $\cos(z_0) = x_0$ and $\sin(z_0) = y_0$.

The change of parameters in a regular curve are also performed via bijective maps between open intervals in $\mathbb{R}$. Here, the statement and the proof of Proposition 1.3.9 can also be naturally adapted to space curves.

Proposition 2.2.7 *Let (I_1, α_1) and (α_2, I_2) be regular parametrizations of the same arc of a regular curve. There exists a bijective map $\gamma : I_1 \to I_2$ with $\gamma \in C^\infty(I_1)$ and $\alpha_2(\gamma(t)) = \alpha_1(t)$, for all $t \in I_1$.*

Proposition 2.2.8 *Given a regular parametrization of an arc of regular curve (I, α), and a one-to-one mapping $\gamma : I_1 \to I_2$ with $\gamma \in C^\infty(I_1)$. Then, the pair $(I_2, \alpha \circ \gamma^{-1})$ is a regular parametrization of the arc.*

We define the velocity vector in the same way as we did for plane curves (see Definition 1.3.11). However, the concept of normal unit vector cannot coincide with that one.

Definition 2.2.9 Let (I, α) be a regular parametrization. The **velocity vector** associated to (I, α) is the function $v_\alpha : I \to \mathbb{R}$ defined by $v_\alpha(t) = \|\alpha'(t)\|$, for $t \in I$.

The vector $T_\alpha(t) = \alpha'(t)/\|\alpha'(t)\|$ is the **tangent unit vector** associated to (I, α) at the point $\alpha(t)$.

Taking into account the definition of the normal vector for regular plane curves (see Definition 1.3.11), one realizes that the adaptation of this definition is not possible when dealing with space curves. This definition relies on the concept of natural parametrizations, which we introduce for the first time in this context. More precisely, the definition of an inflection point in Definition 1.3.16, given in terms of a natural parametrization, describes the curvature in terms of the vector $\alpha''(t)$. The normal vector for plane curves is directly related to the previous one (see (1.12)). This leads to the following definition of the normal vector for space curves.

Definition 2.2.10 A regular parametrization (I, α) is a **natural parametrization** (or an **arc length parametrization**) if $\|\alpha'(t)\| = 1$, for all $t \in I$.

Any regular parametrization can be reparametrized via a natural parametrization, with the same change of parameter as that in the proof of Proposition 1.3.14.

Proposition 2.2.11 *For every regular parametrization (I, α), there exists a natural parametrization of $\alpha(I)$. A change of parameter leading to a natural parametrization is determined by*

$$\gamma^{-1} : \gamma(I) \to I, \quad \gamma(t) = \int_{t_0}^{t} \|\alpha'(u)\| \, du.$$

The computation of the arc length of a curve can be expressed in the same manner as for plane curves. The proof can also be adapted to this framework, arriving at the value of the arc length by means of infinite approximations by segments (see Fig. 2.9).

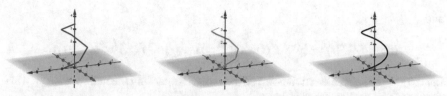

Fig. 2.9 Approximation of a curve with polygonal chains

Proposition 2.2.12 *Let (I, α) be a regular parametrization, and let $t_0 \in I$. For every $t \in I$, the **arc length** between $\alpha(t_0)$ and $\alpha(t)$ is*

$$\int_{t_0}^{t} \left\| \alpha'(u) \right\| du,$$

whose value does not depend on the choice of a reparametrization of the same arc via regular parametrizations.

Lemma 2.2.13 *Let (I, α) be a natural parametrization. Then the vectors $T_\alpha(t)$ and $T_\alpha'(t)$ are orthogonal.*

Proof Taking derivatives at the equality $\langle T_\alpha(t), T_\alpha(t) \rangle = 1$, we get $2 \langle T_\alpha(t), T_\alpha'(t) \rangle = 2 \langle \alpha'(t), \alpha''(t) \rangle = 0$.

The previous result leads to the notion of inflection point in case $T_\alpha'(t) \equiv 0$. Further, we have achieved a procedure to obtain a nonzero vector, which is orthogonal to $T_\alpha(t)$, leading to the concept of normal unit vector.

Definition 2.2.14 Let (I, α) be a natural parametrization, and $t_0 \in I$. We say that $P = \alpha(t_0)$ is an **inflection point** of $\alpha(I)$ if $\alpha''(t_0) = 0$.

In the case where $\alpha(t_0)$ is not an inflection point, we define the **osculating plane** of the curve by the affine plane at $\alpha(t_0)$ associated to the linear span of $\{\alpha'(t_0), \alpha''(t_0)\}$.

Note that the definition of the osculating plane always makes sense when the point is not an inflection point because Proposition 2.2.13 guarantees that $\alpha'(t)$ and $\alpha''(t)$ are orthogonal vectors. As both are not null, they are linearly independent.

The previous definition, stated for normal parametrizations, can be generalized to any regular parametrization of a curve, taking into account Proposition 2.2.11. Observe that, given a regular parametrization of a curve, (I, α) there exists a change of parameter $\gamma : I \to \gamma(I)$ such that $(\gamma(I), \alpha \circ \gamma^{-1})$ is a natural parametrization. It holds that

$$(\alpha \circ \gamma^{-1})'(t) = (\gamma^{-1})'(t)\alpha'(\gamma^{-1}(t)),$$

and

$$(\alpha \circ \gamma^{-1})''(t) = (\gamma^{-1})''(t)\alpha'(\gamma^{-1}(t)) + ((\gamma^{-1})'(t))^2\alpha''(\gamma^{-1}(t)).$$

Let $\alpha \circ \gamma^{-1}(t)$ be an inflection point of the curve. Regarding the natural parametrization $(\gamma(I), \alpha \circ \gamma^{-1})$, we then have

$$0 = (\alpha \circ \gamma^{-1})''(t) = (\gamma^{-1})''(t)\alpha'(\gamma^{-1}(t)) + ((\gamma^{-1})'(t))^2\alpha''(\gamma^{-1}(t)),$$

which means that

$$\frac{-(\gamma^{-1})''(t)}{((\gamma^{-1})'(t))^2}\alpha'(\gamma^{-1}(t)) = \alpha''(\gamma^{-1}(t)).$$

In terms of the parameter $s = \gamma^{-1}(t)$, which is the parameter in which the parametrization $\alpha : I \to \alpha(I)$ is defined, we have

$$\lambda(s)\alpha'(s) = \alpha''(s), \quad \lambda(s) = \frac{-(\gamma^{-1})''(\gamma(s))}{((\gamma^{-1})'(\gamma(s)))^2},$$

i.e., the vectors $\alpha'(s)$ and $\alpha''(s)$ are linearly dependent.

This last statement gives rise to the following definition.

Definition 2.2.15 Let (I, α) be a regular parametrization and $t_0 \in I$. We say $P = \alpha(t_0)$ is an **inflection point** of $\alpha(I)$ if the subspace of $\mathbb{R}^3$ generated by the vectors $\{\alpha'(t_0), \alpha''(t_0)\}$ is of dimension 1, i.e., if $\alpha'(t_0) = \lambda\alpha''(t_0)$ for some $\lambda \in \mathbb{R}$.

In the case where $P = \alpha(t_0)$ is not an inflection point, we define the **osculating plane** of the curve at P by the affine plane at $\alpha(t_0)$ associated to the linear span of $\{\alpha'(t_0), \alpha''(t_0)\}$.

At this point, we are in a position to give a definition of normal unit vector and curvature of a curve at a point starting from a natural parametrization.

Definition 2.2.16 Let (I, α) be a natural parametrization, and $\alpha(t)$ is not an inflection point of $\alpha(I)$. We define the **normal unit vector** of $\alpha(I)$ at $\alpha(t)$ by the unitary vector which is proportional to $\alpha''(t)$, and their inner product is positive. In other words,

$$\alpha''(t) = \kappa N_\alpha(t), \text{ for some } \kappa > 0. \tag{2.8}$$

This definition is equivalent to

$$N_\alpha(t) := \frac{\alpha''(t)}{\|\alpha''(t)\|}.$$

Observe that the expression defining the curvature of a plane curve at a point (see (1.12)) is analogous to (2.8), and leads to the choice of κ as the curvature of a curve at a point.

Definition 2.2.17 Let (I, α) be a natural parametrization, and let $\alpha(t)$ be a point which is not an inflection point of $\alpha(I)$. We define the **curvature** of $\alpha(I)$ at $\alpha(t)$ by the number $\kappa_\alpha(t)$ such that

$$\alpha''(t) = \kappa_\alpha(t) N_\alpha(t). \tag{2.9}$$

Observe that the curvature at a point is always positive, in contrast to the curvature of plane curves (see Definition 1.3.16).

In order to define the curvature of curves described by the image of a regular parametrization, we associate the curvature to the point as follows.

Definition 2.2.18 Let (I, α) be a regular parametrization. We define $\kappa_\alpha(t)$ for all $t \in I$ by $\kappa_\alpha(t) = \kappa_\beta(\gamma(t))$, where $\alpha = \beta \circ \gamma$ and β is a natural parametrization of the same arc of curve.

It will be observed form the next results that the previous definition does not depend on the natural parametrization chosen.

Proposition 2.2.19 *Let (I_1, β_1) and (I_2, β_2) be two natural parametrizations of the same arc. Then there exists $c \in \mathbb{R}$ such that $\beta_1(t) = \beta_2(\pm t + c)$ for all $t \in I_1$, and the intervals I_1 and I_2 are related accordingly.*

The proof in analogous to that of Proposition 1.3.15.

Corollary 2.2.20 *Let (I, α) be a regular parametrization and let (I_1, β_1) and (I_2, β_2) be two natural parametrizations of the same arc of curve. Assume that*

$$\beta_2 \circ \gamma_2 = \beta_1 \circ \gamma_1 = \alpha \tag{2.10}$$

for some changes of parameters $\gamma_1 : I \to I_1$ and $\gamma_2 : I \to I_2$. Then it holds that $\kappa_{\beta_1}(\gamma_1(t)) = \kappa_{\beta_2}(\gamma_2(t))$.

Proof From Proposition 2.2.19 one has

$$\beta_1(t) = \beta_2(t + c) \text{ or } \beta_1(t) = \beta_2(-t + c) \text{ for some } c \in \mathbb{R}. \tag{2.11}$$

It holds that $\|\beta_1(t)\| = \|\beta_2(t + c)\|$, therefore $\kappa_{\beta_1}(t) = \kappa_{\beta_2}(\pm t + c)$, and $\kappa_{\beta_1}(\gamma_1(\tilde{t})) = \kappa_{\beta_2}(\pm\gamma_1(\tilde{t}) + c) = \kappa_{\beta_2}(\gamma_2(\tilde{t}))$, in view of (2.10) and (2.11).

The effective description of the curvature of a regular curve can be obtained as follows.

Proposition 2.2.21 *Let (I, α) be a regular parametrization. Let $\alpha(t)$ be a point in the regular curve which is not an inflection point. Then one has*

$$\kappa_\alpha(t) = \frac{\|\alpha'(t) \times \alpha''(t)\|}{\|\alpha'(t)\|^3}.$$

Proof Proposition 2.2.11 guarantees the existence of a change of parameter γ : $I \to I_1$ with $\alpha = \beta \circ \gamma$, and (I_1, β) is a natural parametrization of the same arc. Let $\alpha(t) = \beta(\gamma(t)) = \beta(s)$ be a point in the curve which is not an inflection point. Regarding Definition 2.2.18 we have

$$\alpha'(t) = \gamma'(t)\beta'(\gamma(t)) = \gamma'(t)T_\beta(\gamma(t)).$$

$$\alpha''(t) = \gamma''(t)T_\beta(\gamma(t)) + (\gamma'(t))^2 T'_\beta(\gamma(t)) = \gamma''(t)T_\beta(\gamma(t)) + (\gamma'(t))^2 \kappa_\beta(\gamma(t))N_\beta(\gamma(t))$$

$$= \gamma''(t)T_\beta(\gamma(t)) + (\gamma'(t))^2 \kappa_\alpha(t)N_\beta(\gamma(t)). \qquad (2.12)$$

The properties of the cross product applied to the previous expression yield

$$\alpha'(t) \times \alpha''(t) = (\gamma'(t))^3 \kappa_\alpha(t)T_\beta(\gamma(t)) \times N_\beta(\gamma(t)).$$

The result follows taking norms in the previous equality.

However, the concept of curvature for plane curves is essentially generalized by that of Definition 2.2.18. More precisely, assume that (I, α) is a regular parametrization of an arc in $\mathbb{R}^2$. We write $\alpha(t) = (\alpha_1(t), \alpha_2(t))$, for all $t \in I$. Then it holds that the parametrization (I, β), with $\beta(t) = (\alpha_1(t), \alpha_2(t), 0)$ is a regular parametrization of a space curve in $\mathbb{R}^3$, contained in the plane $\{z = 0\}$, where both curves coincide. Regarding Propositions 2.2.21 and 1.3.20, it is straightforward to verify that

$$\kappa_\beta(t) = \frac{\left\| \begin{matrix} i & j & k \\ \alpha'_1(t) & \alpha'_2(t) & 0 \\ \alpha''_1(t) & \alpha''_2(t) & 0 \end{matrix} \right\|}{((\alpha'_1(t))^2 + (\alpha'_2(t))^2)^{3/2}} = \frac{|\alpha'_1(t)\alpha''_2(t) - \alpha'_2(t)\alpha''_1(t)|}{((\alpha'_1(t))^2 + (\alpha'_2(t))^2)^{3/2}} = |\kappa_\alpha(t)|.$$

So far, given a natural parametrization (I, α), for every $\alpha(t) \in \alpha(I)$ which is not an inflection point we have defined the tangent unit and normal vectors $T_\alpha(t)$ and $N_\alpha(t)$. Both are orthogonal (see Lemma 2.2.13) and of norm 1. The completion of a basis of $\mathbb{R}^3$ is completed with the unit **binormal vector**.

Fig. 2.10   Frenet trihedron in a space curve

Definition 2.2.22 Let (I, α) be a natural parametrization, and let $\alpha(t) \in \alpha(I)$ be a point which is not an inflection point. We define the corresponding binormal unit vector by

$$B_\alpha(t) = T_\alpha(t) \times N_\alpha(t).$$

Definition 2.2.23 Let (I, α) be a natural parametrization, and let $\alpha(t) \in \alpha(I)$ be a point which is not an inflection point. The set

$$\{\alpha(t); \{T_\alpha(t), N_\alpha(t), B_\alpha(t)\}\}$$

is an affine reference of $\mathbb{R}^3$, known as the **Frenet trihedron**.

Observe that the basis $\{T_\alpha(t), N_\alpha(t), B_\alpha(t)\}$ of the Euclidean space $\mathbb{R}^3$ is orthonormal, in view of the definitions of the vectors involved and the properties of the cross product. It is not possible to define Frenet trihedron in an inflection point of the curve (Fig. 2.10).

Definition 2.2.24 Let (I, α) be a natural parametrization, and let $\alpha(t) \in \alpha(I)$ be a point which is not an inflection point.

The **normal line** to the curve at $\alpha(t)$ is the line at $\alpha(t)$ and director vector $N_\alpha(t)$. The **binormal line** to the curve at $\alpha(t)$ is the line at $\alpha(t)$ and director vector $B_\alpha(t)$.

The **osculating plane** related to the curve at $\alpha(t)$ is the plane at $\alpha(t)$ and associated vector space generated by $\{T_\alpha(t), N_\alpha(t)\}$. The **normal plane** related to the curve at $\alpha(t)$ is the plane at $\alpha(t)$ and associated vector space generated by $\{N_\alpha(t), B_\alpha(t)\}$. The **rectifying plane** related to the curve at $\alpha(t)$ is the plane at $\alpha(t)$ and associated vector space generated by $\{T_\alpha(t), B_\alpha(t)\}$ (Fig. 2.11).

The Frenet trihedron has been defined in association to normal parametrizations. Taking into account Proposition 2.2.11, given a regular parametrization (I, α), there exists a change of parameter $\gamma : I_1 \rightarrow I$ such that $(I_1, \alpha \circ \gamma)$ is a natural parametrization of the same arc of curve. Definition 2.2.15 characterizes the points $\alpha(t)$ for which the Frenet trihedron is well defined, and consequently the lines and planes associated to it. This leads to the next definition.

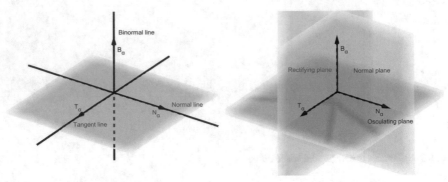

Fig. 2.11 Lines and planes associated to the Frenet trihedron

Definition 2.2.25 Let (I, α) be a regular parametrization. Let $\alpha(t) \in \alpha(I)$ be a point which is not an inflection point of the curve. We define

$$T_\alpha(t) = \frac{\alpha'(t)}{\|\alpha'(t)\|},$$
$$(2.13)$$

$$B_\alpha(t) = \frac{\alpha'(t) \times \alpha''(t)}{\|\alpha'(t) \times \alpha''(t)\|},$$
$$(2.14)$$

$$N_\alpha(t) = B_\alpha(t) \times T_\alpha(t).$$
$$(2.15)$$

Observe that the previous definition extends that of the Frenet trihedron when (I, α) is a normal parametrization because in that case $T_\alpha(t) = \alpha'(t)$. In accordance with (2.13), the normal vector is in the direction of $\alpha''(t)$, so the binormal vector is given by $B_\alpha(t) = \alpha'(t) \times N_\alpha(t)$, which coincides with (2.14). Finally, observe that the vector in (2.15) is chosen among the two possible choices of orthonormal vectors to the previous ones. Its construction shows that the direction chosen coincides with that of natural parametrizations. In addition to that, all these three vectors are unitary, in view of their definition.

In contrast to plane curves, whose information is stored essentially in the curvature (see Theorem 1.3.22), space curves need to be specified by giving more information. The curvature at a point of a normal parametrization describes how much the tangent vector varies in the vicinity points of the curve. In fact,

$$\kappa_\alpha(t) = \frac{1}{\|\alpha''(t)\|}.$$

If the point is not an inflection point:

$$T'_\alpha(t) = \kappa_\alpha(t) N_\alpha(t).$$
$$(2.16)$$

Let (I, α) be a natural parametrization, and let $\alpha(t)$ be a point which is not an inflection point of the curve. We have $1 = \langle B_\alpha(t), B_\alpha(t) \rangle$. Taking derivatives in the previous expression we find that $B'_\alpha(t)$ is orthogonal to $B_\alpha(t)$. Moreover, the definition of the binormal unit vector yields

$$B'_\alpha(t) = T'_\alpha(t) \times N_\alpha(t) + T_\alpha(t) \times N'_\alpha(t)$$
$$= \kappa_\alpha(t) N_\alpha(t) \times N_\alpha(t) + T_\alpha(t) \times N'_\alpha(t) = T_\alpha(t) \times N'_\alpha(t),$$

so $B'_\alpha(t)$ is also orthogonal to $T_\alpha(t)$ (and $N'_\alpha(t)$). Therefore, $B'_\alpha(t)$ and $N_\alpha(t)$ are linearly dependent vectors. This leads to the following definition.

Definition 2.2.26 Let (I, α) be a natural parametrization, such that $\alpha(t)$ is not an inflection point of $\alpha(I)$. We define the torsion of the curve at $\alpha(t)$, and denote it by $\tau_\alpha(t)$, by

$$B'_\alpha(t) = -\tau_\alpha(t) N_\alpha(t). \tag{2.17}$$

Observe that $\tau_\alpha(t)$ can be positive or negative, in contrast to the curvature of a curve at a point. Equation (2.17) can also be interpreted from the same point of view as the curvature in (2.16), so the torsion at a point in a normal parametrization describes how much the binormal vector varies in the vicinity points of the curve. The sign of the torsion at a point shows the way this variation holds. An illustrative example will help us understand better.

Example 2.2.27 Let $(\mathbb{R}, \alpha)$ be the regular parametrization defined by

$$\alpha(t) = (a\cos(t), a\sin(t), bt),$$

for some $a > 0$ and $b \in \mathbb{R}$. Assume that $a^2 + b^2 = 1$ so the parametrization is natural.

We have that

$$\alpha'(t) = (-a\sin(t), a\cos(t), b), \quad t \in \mathbb{R}.$$

$$\alpha''(t) = (-a\cos(t), -a\sin(t), 0), \quad t \in \mathbb{R},$$

arriving at

$$T_\alpha(t) = (-a\sin(t), a\cos(t), b),$$
$$N_\alpha(t) = (-\cos(t), -\sin(t), 0),$$
$$B_\alpha(t) = (b\sin(t), -b\cos(t), a), \tag{2.18}$$

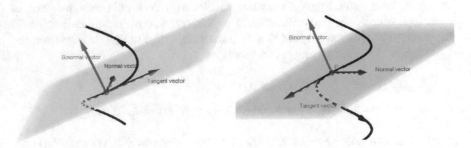

Fig. 2.12 Positive torsion (left) and negative torsion (right) in Example 2.2.27

for all $t \in \mathbb{R}$. Equation (2.17) allows us to determine that $\tau_\alpha(t) = b$. Observe in Fig. 2.12 how the curve separates the osculating plane in the case of a positive and negative torsion.

Example 2.2.28 We consider the loxodrome parametrized by (2.2), with $n = 2$ and $R = r = 1$. Definition 2.2.25 can be applied here to come up with the elements of the Frenet trihedron. We have, for all $t \in \mathbb{R}$,

$$T_\alpha(t) = \left(\frac{-r \sin(t)(2 \cos(t) + 1)}{r\sqrt{1 + (\cos(t) + 1)^2}}, \frac{r(\cos(t) + 2\cos(t)^2 - 1)}{r\sqrt{1 + (\cos(t) + 1)^2}}, \frac{r \cos(t)}{r\sqrt{1 + (\cos(t) + 1)^2}} \right),$$

$$N_\alpha(t) = B_\alpha(t) \times T_\alpha(t),$$

$$B_\alpha(t) = \left(\frac{r^2 \sin(t)(2 \cos(t)^2 + 1)}{r^2\sqrt{11 + 18\cos(t) + 4\cos(t)^3 + 12\cos(t)^2}}, \right.$$

$$\frac{r^2(2\cos(t)^3 + 1)}{r^2\sqrt{11 + 18\cos(t) + 4\cos(t)^3 + 12\cos(t)^2}},$$

$$\left. \frac{3r^2(1 + \cos(t))}{r^2\sqrt{11 + 18\cos(t) + 4\cos(t)^3 + 12\cos(t)^2}} \right). \qquad (2.19)$$

Theorem 2.2.29 *Let (I, α) be a natural parametrization of a curve without inflection points. Then the following statements are equivalent:*

- *$\alpha(I)$ is a plane curve.*
- *$\tau_\alpha(t) = 0$ for all $t \in I$.*

Proof First, assume the curve is contained in a plane. Its osculating plane does not vary with $t \in I$, so the binormal unit vector is constant, and takes only one value for all $t \in I$ because I is a connected set. Then $B'_\alpha(t) = 0 = -\tau_\alpha(t)N_\alpha(t)$, so $\tau_\alpha(t)$ vanishes for all $t \in I$. Second, if $\tau_\alpha(t) = 0$ for all $t \in I$, then (2.17) yields $B_\alpha(t)$

is a constant vector. Then it holds that $\langle \alpha(t) - \alpha(t_0), B_\alpha(t_0) \rangle \equiv 0$, $\alpha(t)$ is contained in the osculating plane for all $t \in I$, which is constant.

Note that $B_\alpha(t)$ is the normal vector of the osculating plane, so the torsion measures how fast the curve separates from the osculating plane, or in other words, how "plane" the curve is.

Given another natural parametrization of a curve, the value of the torsion associates to both parametrizations only differs in the sign in view of Proposition 2.2.19, so its value is essentially invariant.

There are other interesting properties which can be derived from direct computations. Consider the equality $0 \equiv \langle B_\alpha(t), N_\alpha(t) \rangle$. Taking derivatives of this equality we get that

$$\tau_\alpha(t) = \langle B_\alpha(t), N_\alpha'(t) \rangle = -\langle N_\alpha(t), B_\alpha'(t) \rangle.$$

Lemma 2.2.30 *Let* (I, α) *be a natural parametrization, and* $\kappa_\alpha(t) \neq 0$. *Then it holds that*

$$N_\alpha'(t) = -\kappa_\alpha(t) T_\alpha(t) + \tau_\alpha(t) B_\alpha(t).$$

Proof Observe that $\{T_\alpha(t), N_\alpha(t), B_\alpha(t)\}$ is a basis of $\mathbb{R}^3$ so the vector $N_\alpha'(t)$ can be written as a linear combination of them. Moreover, $N_\alpha'(t)$ is orthogonal to $N_\alpha(t)$ because $\langle N_\alpha(t), N_\alpha(t) \rangle = 1$ and taking derivatives in the previous equality we get $\langle N_\alpha(t), N_\alpha'(t) \rangle = 0$. Due to the fact that they are orthogonal vectors, the linear combination does not involve $N_\alpha(t)$. In order to find the coefficients multiplying $T_\alpha(t)$ and $B_\alpha(t)$ we apply the properties of the Fourier coefficients in an Euclidean space. The coefficient multiplying $T_\alpha(t)$ in the linear combination is given by $\langle T_\alpha(t), N_\alpha'(t) \rangle$. Taking derivatives at $\langle T_\alpha(t), N_\alpha(t) \rangle \equiv 0$ and from (2.16) we obtain that the coefficient is

$$\langle T_\alpha(t), N_\alpha'(t) \rangle = -\langle T_\alpha'(t), N_\alpha(t) \rangle = -\kappa_\alpha(t) \langle N_\alpha(t), N_\alpha(t) \rangle = -\kappa_\alpha(t).$$

The coefficient multiplying $B_\alpha(t)$ can be found in a similar manner, using $\langle B_\alpha(t), N_\alpha(t) \rangle$ and (2.17).

Proposition 2.2.31 *Let* (I, α) *be a natural parametrization, and* $\kappa_\alpha(t) \neq 0$. *Then it holds that*

$$\tau_\alpha(t) = \frac{[\alpha'(t), \alpha''(t), \alpha'''(t)]}{\langle \alpha''(t), \alpha''(t) \rangle}.$$

Proof We have $\alpha'(t) = T_\alpha(t)$, and from (2.16) we have $\alpha''(t) = \kappa_\alpha(t) N_\alpha(t)$. Taking derivatives in the previous expression yields

$$\alpha'''(t) = (\kappa_\alpha(t) N_\alpha(t))' = \kappa_\alpha'(t) N_\alpha(t) + \kappa_\alpha(t) N_\alpha'(t).$$

Regarding Lemma 2.2.30, we obtain that

$$\alpha'''(t) = \kappa_\alpha'(t)N_\alpha(t) + \kappa_\alpha(t)\left(-\kappa_\alpha(t)T_\alpha(t) + \tau_\alpha(t)B_\alpha(t)\right).$$

It holds that

$$\alpha''(t) \times \alpha'''(t) = -\kappa_\alpha(t)^3(N_\alpha(t) \times T_\alpha(t)) + \kappa_\alpha(t)^2\tau_\alpha(t)(N_\alpha(t) \times B_\alpha(t)).$$

Therefore, in the computation of $[\alpha'(t), \alpha''(t), \alpha'''(t)] = \alpha'(t)\cdot(\alpha''(t)\times\alpha'''(t))$, only the coefficient $N_\alpha(t) \times B_\alpha(t)$ does not vanish, and indeed $1 = T_\alpha(t)\cdot(N_\alpha(t)\times B_\alpha(t))$, which yields that

$$[\alpha'(t), \alpha''(t), \alpha'''(t)] = \kappa_\alpha^2(t)\tau_\alpha(t).$$

The proof can be concluded by verifying that

$$\alpha''(t) \cdot \alpha''(t) = T_\alpha'(t) \cdot T_\alpha'(t) = \kappa_\alpha(t)^2 N_\alpha(t) \cdot N_\alpha(t) = \kappa_\alpha(t)^2.$$

Regarding regular parametrizations, we define the torsion at a point from the existing relation with natural parametrizations.

Definition 2.2.32 Let (I, α) be a regular parametrization with $\kappa_\alpha(t) \neq 0$. We define $\tau_\alpha(t)$ by $\tau_\alpha(t) = \tau_\beta(\gamma(t))$, where $\alpha = \beta \circ \gamma$ and β is a natural parametrization of the same arc of curve.

Observe from Proposition 2.2.19 that the sign of the torsion can differ depending on the natural parametrization chosen. Also, a formula derived from Definition 2.2.32 and Proposition 2.2.31 can be obtained for regular parametrizations.

Proposition 2.2.33 *Let (I, α) be a regular parametrization with $\kappa_\alpha(t) \neq 0$. It holds that*

$$\tau_\alpha(t) = \frac{[\alpha'(t), \alpha''(t), \alpha'''(t)]}{\|\alpha'(t) \times \alpha''(t)\|^2}.$$

Proof Let γ be a change of parameter such that $\alpha = \beta \circ \gamma$ with β being a natural parametrization of $\alpha(I)$. For the sake of simplicity, we assume that $\gamma'(t) > 0$ for all t. One has

$$\tau_\alpha(t) = \tau_\beta(\gamma(t)) = \tau_\beta(s) = \frac{[\beta'(s), \beta''(s), \beta'''(s)]}{\langle\beta''(s), \beta''(s)\rangle}.$$

On the one hand, $\alpha'(t) = \gamma'(t)\beta'(s)$,

$$\alpha''(t) = (\gamma'(t))^2\beta''(s) + \gamma''(t)\beta'(s),$$

$$\alpha'''(t) = (\gamma'(t))^3\beta'''(s) + 3\gamma'(t)\gamma''(t)\beta''(s) + \gamma'''(t)\beta'(s),$$

which yields

$$\alpha'(t) \cdot (\alpha''(t) \times \alpha'''(t)) = (\gamma'(t))^6 \beta'(s) \cdot (\beta''(s) \times \beta'''(s)) = (\gamma'(t))^6 [\beta'(s), \beta''(s), \beta'''(s)].$$

Moreover, regarding the proof of Proposition 2.2.31, we get

$$[\beta'(s), \beta''(s), \beta'''(s)] = \kappa_\beta^2(s) \tau_\beta(s). \tag{2.20}$$

On the other hand,

$$\left\| \alpha'(t) \times \alpha''(t) \right\|^2 = \left\| (\gamma'(t))^3 (T_\beta(s) \times \kappa_\beta(s) N_\beta(s)) \right\|^2 = |\gamma'(t)|^6 \kappa_\alpha^2(t). \tag{2.21}$$

Regarding (2.20) and (2.21) we conclude the result.

Taylor's expansion and the expression of the Frenet trihedron lead to the asymptotic behavior of a curve near a point under study. Indeed, given a natural parametrization (I, α), let $\alpha(t_0) \in \alpha(I)$ such that $\kappa_\alpha(t_0) \neq 0$. We have that $\alpha(t)$ behaves in the vicinity of $\alpha(t_0)$ as

$$\alpha(t) = \alpha(t_0) + \alpha'(t_0)(t - t_0) + \frac{\alpha''(t_0)}{2!}(t - t_0)^2 + \frac{\alpha'''(t_0)}{3!}(t - t_0)^3 + \ldots$$

Regarding the expressions of $\alpha'(t_0)$, $\alpha''(t_0)$ and $\alpha'''(t_0)$ in the proof of Proposition 2.2.31, we obtain that

$$\alpha(t) = \alpha(t_0) + \left((t - t_0) - \frac{\kappa_\alpha^2(t_0)}{2}(t - t_0)^3 \right) T_\alpha(t_0)$$

$$+ \left(\frac{\kappa_\alpha(t_0)}{2}(t - t_0)^2 + \frac{\kappa_\alpha'(t_0)}{6}(t - t_0)^3 \right) N_\alpha(t_0)$$

$$+ \left(\frac{\kappa_\alpha(t_0)\tau_\alpha(t_0)}{6}(t - t_0)^3 \right) B_\alpha(t_0) + \ldots \tag{2.22}$$

In the reference determined by the vectors $\{T_\alpha(t_0), N_\alpha(t_0), B_\alpha(t_0)\}$ we find that the curve behaves like the plane curve with parametrization given by

$$x(t) = t - t_0, \qquad y(t) = \frac{\kappa_\alpha(t_0)}{2}(t - t_0)^2$$

near $\alpha(t_0)$, and in the projection on the plane $\{T_\alpha(t_0), N_\alpha(t_0)\}$;

$$x(t) = t - t_0, \qquad y(t) = \frac{\kappa_\alpha(t_0)\tau_\alpha(t_0)}{6}(t - t_0)^3$$

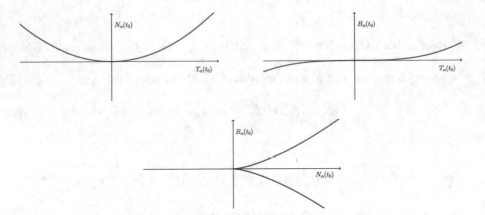

Fig. 2.13 Local shape of a curve at a point

near $\alpha(t_0)$, in the projection on the plane $\{T_\alpha(t_0),\ B_\alpha(t_0)\}$;

$$x(t) = \frac{\kappa_\alpha(t_0)}{2}(t - t_0)^2, \qquad y(t) = \frac{\kappa_\alpha(t_0)\tau_\alpha(t_0)}{6}(t - t_0)^3$$

near $\alpha(t_0)$, in the projection on the plane $\{N_\alpha(t_0),\ B_\alpha(t_0)\}$. These curves are illustrated in Fig. 2.13 in the case of $t_0 = 0$ and $\kappa_\alpha(t_0) = \tau_\alpha(t_0) = 1$.

We consider the natural parametrization $(\mathbb{R}, \alpha)$ given by

$$\alpha(t) = (a\cos(t),\ a\sin(t),\ bt), \quad t \in \mathbb{R}$$

where a, b are positive parameters with $a^2 + b^2 = 1$. In Fig. 2.14 we display the shape of the curve in the vicinity of a point, which is coherent with the projections described above.

The concepts of curvature (see (2.16)), torsion (see (2.17)), and Lemma 2.2.30 lead to the so-called **Frenet formulas**.

Theorem 2.2.34 *Given a natural parametrization (I, α), with $\kappa_\alpha(t) \neq 0$ for all $t \in I$, then the Frenet formulas hold:*

$$\begin{cases} T_\alpha'(t) = & \kappa_\alpha(t)N_\alpha(t) \\ N_\alpha'(t) = -\kappa_\alpha(t)T_\alpha(t) + & + \tau_\alpha(t)B_\alpha(t). \\ B_\alpha'(t) = & -\tau_\alpha(t)N_\alpha(t) \end{cases}$$

These relations can be rewritten in the case where (I, α) is a regular parametrization with the help of a reparametrization.

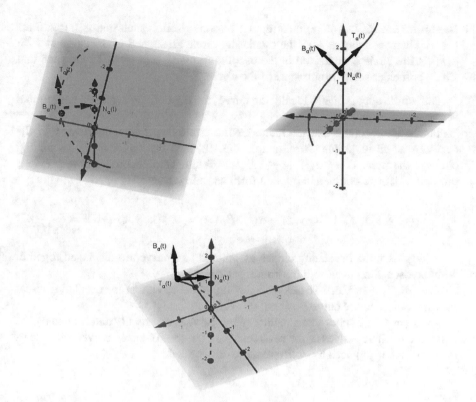

Fig. 2.14 Example of local shape of a curve at a point

Corollary 2.2.35 *Given a regular parametrization* (I_1, β), *with* $\kappa_\beta(s) \neq 0$ *for all* $s \in I_1$, *then the Frenet formulas hold:*

$$\begin{cases} T'_\beta(s) = & \|\beta'(s)\| \,\kappa_\beta(s) N_\beta(s) \\ N'_\beta(s) = - \|\beta'(s)\| \,\kappa_\beta(s) T_\beta(s) + & + \|\beta'(s)\| \,\tau_\beta(s) B_\beta(s) \\ B'_\beta(s) = & - \|\beta'(s)\| \,\tau_\beta(s) N_\beta(s) \end{cases}$$

Proof Let $\gamma : I \rightarrow I_1$ be a change of parameter with $\gamma'(t) > 0$, such that $\alpha = \beta \circ \gamma$ is a natural parametrization of $\beta(I_1)$. It is straightforward to verify that $T_\alpha(t) = T_\beta(\gamma(t))$, $N_\alpha(t) = N_\beta(\gamma(t))$ and $B_\alpha(t) = B_\beta(\gamma(t))$. Taking into account that $\kappa_\alpha(t) = \kappa_\beta(\gamma(t))$ and $\tau_\alpha(t) = \tau_\beta(\gamma(t))$, the result follows.

An interesting construction is that of the **osculating circle** associated to a regular curve.

Definition 2.2.36 Let (I, α) be a regular parametrization such that $\kappa_\alpha(t) \neq 0$ for $t \in I$. Then for every $t_0 \in I$, the osculating circle associated to the curve at $P = \alpha(t_0)$ is the circle C contained in the osculating plane, with $P \in C$ and such that $\kappa_\alpha(t_0)$ coincides with the curvature of the circle.

That circle approximates locally the curve, having both unit tangent and normal vector in common.

Taking into account that the radius of the circle determines the inverse of the curvature at all its points (see Example 1.3.19), then the center of the osculating circle is the point $C - \alpha(t) + \frac{1}{\kappa_\alpha(t)} N_\alpha(t)$. Taking into account all the previous properties, it is determined in implicit form as follows:

$$C = \{(x, y, z) \in \mathbb{R}^3 : \langle (x, y, z) - \alpha(t), B_\alpha(t) \rangle = 0, \|(x, y, z) - C\| = \frac{1}{\kappa_\alpha(t)}\}.$$

In Fig. 2.15, the osculating circles associated to a curve are displayed together with the osculating plane and the Frenet trihedron.

The QR Code of Fig. 2.16 links to a display of the motion of the osculating circle moving along a space curve.

Theorem 1.3.22 states that a regular plane curve is essentially determined by the curvature at all the values of the parameters. In the case of space curves, the role of the curvature is replaced by both curvature and torsion.

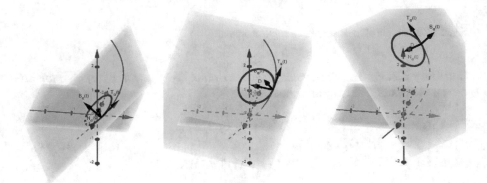

Fig. 2.15 Some osculating circles associated to a curve

Fig. 2.16 QR Code 11

Theorem 2.2.37 (Fundamental Theorem of Space Curves) *Let* $\kappa, \tau : I \to \mathbb{R}$ *be in* $C^\infty(I)$, *with* $\kappa(t) > 0$ *for every* $t \in I$, *for some nonempty open interval* $I \subseteq \mathbb{R}$. *Then there exists a natural parametrization* (I, α) *such that* $\kappa(t) = \kappa_\alpha(t)$ *and* $\tau(t) = \tau_\alpha(t)$ *for all* $t \in I$. *Any other regular curve under this property results from a rigid transformation of the first curve in* $\mathbb{R}^3$ *(an affine transformation associated to an isometry).*

It is worth mentioning that the curve obtained by a rigid transformation of the first one can be parametrized in different ways, but the curve is always reparametrizable via a natural parametrization. Therefore, given any other curve which is parametrized by a natural parametrization $(J, \tilde{\alpha})$, it holds that $\alpha(t) = \Phi(\tilde{\alpha}(t))$, where Φ is an affine transformation associated to an isometry in $\mathbb{R}^3$. Section 2.4 gives more detail on the parametric formulation of a rigid transformation, with reference to some architectural elements.

Moreover, observe that in the case of plane curves, Theorem 2.2.29 states that the function $\tau \equiv 0$ for both the initial curve and that after any rigid transformation. This means that the rigid transformation can be restricted to the osculating plane, where both curves are contained. That rigid transformation is a two-dimensional one.

Outline of the Proof The existence of the curve in the result is guaranteed by the existence of solutions of a Cauchy problem determined by Frenet equations in Corollary 2.2.35. Note that we start from a linear system of differential equations of first order

$$\frac{d}{dt}\begin{pmatrix} T_\alpha(t) \\ N_\alpha(t) \\ B_\alpha(t) \end{pmatrix} = \begin{pmatrix} 0 & \kappa_\alpha(t) & 0 \\ -\kappa_\alpha(t) & 0 & \tau_\alpha(t) \\ 0 & -\tau_\alpha(t) & 0 \end{pmatrix} \begin{pmatrix} T_\alpha(t) \\ N_\alpha(t) \\ B_\alpha(t) \end{pmatrix}, \qquad (2.23)$$

for every $t \in I$. The solutions of this differential system of equations is an orthonormal basis in $\mathbb{R}^3$: $\{T_\alpha(t), N_\alpha(t), B_\alpha(t)\}$.

We define

$$\alpha(t) = \alpha(t_0) + \int_{t_0}^t T_\alpha(s)\,ds, \quad t \in I.$$

which turns out to define the parametrization of a curve whose curvature and torsion are those fixed in the statements of the theorem.

In order to prove unicity of the curve, given any two orthonormal bases

$$\{T_\alpha(t), N_\alpha(t), B_\alpha(t)\} \text{ and } \{\tilde{T}_\alpha(t), \tilde{N}_\alpha(t), \tilde{B}_\alpha(t)\}$$

which solve the system (2.23), we can transform one into the other by means of a rigid transformation in $\mathbb{R}^3$.

Regarding the proof of Theorem 2.2.37, we could, at least theoretically, construct any regular curve with given curvature and torsion at every point. This construction is unique after rigid transformations (see Sect. 2.4). In practice, the solution of the 9×9 system of linear differential equations determined by Frenet equations (see Corollary 2.2.35) could be difficult to solve depending on the choice of the functions $\kappa(t)$ and $\tau(t)$. In addition to this, the curve is defined by integration of the tangent vector obtained by solving the previous system, which might not be solvable or valid for practical purposes.

However, in simple examples, one can construct the curve following this procedure. Other situations might be manageable after adopting simpler approximated data.

Example 2.2.38 Let $\kappa(t) = c$ for some $c > 0$, and choose $\tau(t) = 0$ for all $t \in \mathbb{R}$. This choice guarantees that the space curve is indeed a plane curve by Theorem 2.2.29.

The system

$$\frac{d}{dt} \begin{pmatrix} T_\alpha(t) \\ N_\alpha(t) \\ B_\alpha(t) \end{pmatrix} = \begin{pmatrix} 0 & c & 0 \\ -c & 0 & 0 \\ 0 & 0 & 0 \end{pmatrix} \begin{pmatrix} T_\alpha(t) \\ N_\alpha(t) \\ B_\alpha(t) \end{pmatrix},$$

admits

$$T_\alpha(t) = (C_1 \sin(ct) - C_2 \cos(ct), C_3 \sin(ct) - C_4 \cos(ct), C_5 \sin(ct) - C_6 \cos(ct)),$$

$$N_\alpha(t) = (C_2 \sin(ct) + C_1 \cos(ct), C_4 \sin(ct) + C_3 \cos(ct), C_6 \sin(ct) + C_5 \cos(ct)),$$

$$B_\alpha(t) = (C_7, C_8, C_9) \qquad (2.24)$$

as general solution, for $C_1, \ldots, C_9 \in \mathbb{R}$. We consider the Frenet trihedron at $t = 0$ given by $T_\alpha(0) = (1, 0, 0)$, $N_\alpha(0) = (0, 1, 0)$ and $B_\alpha(0) = (0, 0, 1)$. We conclude that $T_\alpha(t) = (\cos(ct), \sin(ct), 0)$, $t \in \mathbb{R}$. We define

$$\alpha(t) = \int_0^t T_\alpha(s)ds = (\frac{1}{c} \sin(ct), \frac{1 - \cos(ct)}{c}, 0),$$

which is a parametrization of the circle centered at $(0, 1/c)$, contained in the floor plane, and radius $1/c$. Note that this is coherent with the results obtained in Example 1.3.19.

Before concluding this chapter, we mention an application of the Frenet trihedron in tiling freeform shapes with straight panels by means of geodesic curves, described in Wallner et al. (2010).

2.3 Some Classic Space Curves in Architecture

This section provides a practical view of the theory presented in the previous sections through architectural elements containing space curves.

In the first part of this chapter, we enumerate some of the classic space curves which appear in or inspire an architectural work. Some of them are described in Sect. 3.4, when dealing with curves associated to surfaces. In Conversano et al. (2011), the authors focus on three types of spatial curves which appear recursively in architecture through history. We also refer to Birindelli and Cedrone (2012). We first focus on the most outstanding and classic space curve:

Helix

Helices are the archetype of space curves. Their appearance in architectural structures is widespread, since other space curves are obtained by adequate modifications of them. Here, we describe their geometric representation in terms of the objects studied above.

A circular helix is a regular curve determined by the parametrization $(\mathbb{R}, \alpha)$, where

$$\alpha(t) = (\cos(t), \sin(t), t), \quad t \in \mathbb{R}. \tag{2.25}$$

Figure 2.17 illustrates a circular helix.

The mathematical elements associated to this curve are described in Sect. 2.5.

The shape of the projection of a helix on the coordinate planes can be observed by choosing the corresponding components in its parametrization. The projection on the XY plane parametrizes a circle centered at the origin of $\mathbb{R}^2$ with unit radius. The projection traverses the circle for any choice of interval of length 2π in the

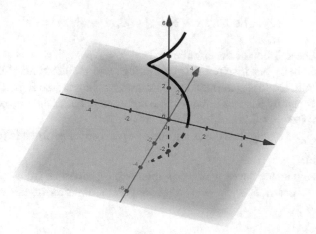

Fig. 2.17 Circular helix

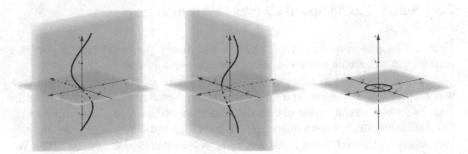

Fig. 2.18 Projections of the circular helix

Fig. 2.19 QR Code 12

parameter. Its projection on the XZ plane describes the graph of cosine function, whilst its projection on the YZ plane draws the graph of sine function.

Observe the three projections of the circular helix in Fig. 2.18.

The QR Code in Fig. 2.19 links to the three different projections of the helix constructing the plane curves observed above.

In view of the parametric definition of a helix in (2.25), it is straightforward to verify that

$$\{(x, y, z) \in \mathbb{R}^3 : x - \cos(z) = 0; \ y - \sin(z) = 0\}$$

is an implicit description of the circular helix.

The classic helix has inspired many buildings and architectural elements, both built and unbuilt. Examples of these are Solomonic columns (see Fig. 2.20, left); the Helix City Project by Kisho Kurokawa; the columns at Park Güell in Barcelona by Antoni Gaudí (see Fig. 2.20, right); and many others.

In Sect. 2.5, we manipulate the definition of a helix to transform it to fit our needs.

Twisted Cubic

This is a classic space curve which has interesting properties from the projective point of view. Let $a, b, c > 0$. A regular parametrization of the curve is the following:

$$\alpha(t) = (at, bt^2, ct^3), \quad t \in \mathbb{R}. \tag{2.26}$$

Fig. 2.20 Examples of helices in architecture

Observe the curve is also defined by

$$C = \{(x, y, z) \in \mathbb{R}^3 : bx^2 - ay = 0; cx^3 - az = 0\}.$$

The elements associated to this curve can be computed directly, obtaining that

$$\tau_\alpha(t) = \frac{3abc}{9b^2c^2t^4 + 9a^2c^2t^2 + a^2b^2}, \quad \kappa_\alpha(t) = 2\sqrt{\frac{9b^2c^2t^4 + 9a^2c^2t^2 + a^2b^2}{(a^2 + 4b^2t^2 + 9c^2t^4)^3}}.$$

From the aesthetic point of view, this space curve is very interesting. Depending on the point from which a viewer perceives the object, one may observe a parabola, a cubic, or a cusp. Observe that the projection of the parametrization in (2.26) draws a parabola projected onto the XY plane, a cubic curve in the XZ plane, and a cusp in the YZ plane (Fig. 2.21). This configuration can be observed in Fig. 2.22.

The phenomena occurred regarding the twisted cubic has been the inspiration for buidings whose shapes differ depending on the point of view considered. Examples of this are the twisted cubic design for the Taipei City Museum of Art by Portuguese architects OODA (Inhabitat 2019c), or the Capital Gate Tower in Abu Dhabi by RMJM (see Fig. 2.23).

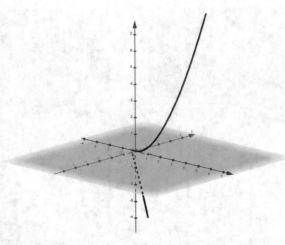

Fig. 2.21 Twisted cubic (2.26), for $a = b = c = 1$

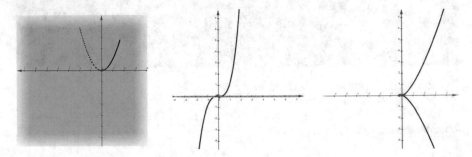

Fig. 2.22 Different projections of the twisted cubic

In Sect. 3.4 we also describe some curves obtained by the intersection of surfaces which usually appear in architectural elements.

2.4   Rigid Transformations in $\mathbb{R}^3$

Theorem 2.2.37 states that any regular space curve is essentially determined by its curvature and torsion at any point, modulo a rigid transformation.

A rigid transformation in $\mathbb{R}^3$ can be defined as a transformation of the whole three-dimensional Euclidean space into itself, $T : \mathbb{R}^3 \to \mathbb{R}^3$, which preserves distances, i.e.,

$$d(P, Q) = d(T(P), T(Q)), \quad P, Q \in \mathbb{R}^3.$$

Fig. 2.23 Capital Gate Tower in Abu Dhabi

or equivalently,

$$\left\| \vec{PQ} \right\| = \left\| \vec{T(P)T(Q)} \right\|, \quad P, Q \in \mathbb{R}^3. \tag{2.27}$$

At first, we might think that such transformations are quite general, involving a wide range of transformations such as exponentials, logarithms, polynomials, etc. However, the fact that the distance is preserved only allows very specific forms of such rigid transformations, which are associated to an homomorphism and a translation.

Lemma 2.4.1 *Let $T : \mathbb{R}^3 \to \mathbb{R}^3$ be a rigid transformation. Then it holds that*

$$T(x, y, z) = (b_1 + a_{11}x + a_{12}y + a_{13}z, \, b_2 + a_{21}x + a_{22}y + a_{23}z, \, b_3 + a_{31}x + a_{32}y + a_{33}z), \tag{2.28}$$

for some $a_{ij}, b_i \in \mathbb{R}$, for all $(x, y, z) \in \mathbb{R}^3$.

Outline of the Proof First assume that $T(O) = O$, where O stands for the origin of coordinates. Then by the definition and properties of the Euclidean distance, it holds that

$$d(T(P + Q), T(P) + T(Q)) = 0, \quad d(T(\lambda P), \lambda T(P)) = 0,$$

for all $P, Q \in \mathbb{R}^3$ and $\lambda \in \mathbb{R}$. Then the map can be represented by a $3x3$ matrix so that $T(x, y, z) = (x', y', z')$, with

$$\begin{pmatrix} x' \\ y' \\ z' \end{pmatrix} = \begin{pmatrix} a_{11} & a_{12} & a_{13} \\ a_{21} & a_{22} & a_{23} \\ a_{31} & a_{32} & a_{33} \end{pmatrix} \begin{pmatrix} x \\ y \\ z \end{pmatrix}.$$

In case $T(O) = (b_1, b_2, b_3)$, then T composed with the traslation $Tr(x, y, z) = (x - b_1, y - b_2, z - b_3)$ is a rigid transformation of the first type, so it is represented in the form of (2.28).

Observe from the proof that any rigid transformation can in fact be interpreted by a translation together with the action of a linear map from the vector space $\mathbb{R}^3$ to $\mathbb{R}^3$. In addition to this, its matrix representation is given by

$$\begin{pmatrix} 1 \\ x' \\ y' \\ z' \end{pmatrix} = \begin{pmatrix} 1 & 0 & 0 & 0 \\ b_1 & a_{11} & a_{12} & a_{13} \\ b_2 & a_{21} & a_{22} & a_{23} \\ b_3 & a_{31} & a_{32} & a_{33} \end{pmatrix} \begin{pmatrix} 1 \\ x \\ y \\ z \end{pmatrix}, \tag{2.29}$$

where (b_1, b_2, b_3) determines the translation $Tr : \mathbb{R}^3 \rightarrow \mathbb{R}^3$

$$Tr(x, y, z) = (x + b_1, y + b_2, z + b_3), \quad (x, y, z) \in \mathbb{R}^3.$$

Not every transformation of the form (2.28) (or equivalently (2.29)) is a rigid transformation, but any rigid transformation is of that form. Indeed, in order for (2.28) to be a rigid transformation, then the linear map from $\mathbb{R}^3$ to $\mathbb{R}^3$, which in the canonical basis is represented by the matrix

$$A = \begin{pmatrix} a_{11} & a_{12} & a_{13} \\ a_{21} & a_{22} & a_{23} \\ a_{31} & a_{32} & a_{33} \end{pmatrix},$$

turns out to be a linear isometry.

The theory needed to classify and construct all the isometries in $\mathbb{R}^3$ makes use of the theory of eigenvalues and eigenvectors of a square matrix so we omit the details at this point, and refer to Chapter 12 in Lang (1986) for further details.

Lemma 2.4.2 *Let* $A \in \mathcal{M}_{n \times n}(\mathbb{R})$. *A represents an isometry if and only if* $AA^T = I$.

Observe that A is invertible, with $\det(A) = \pm 1$. After an orthonormal change of coordinates that does not involve any modification on the distances but a different "point of view" of the objects in $\mathbb{R}^3$, one has the following rigid transformations:

- Rotation around a line.
- Reflection with respect to a plane.

Any other transformation is a composition of the previous ones. In particular, if the rotation is of angle $\alpha = 0$, we obtain the identity, and in case it is of angle $\alpha = \pi$, one gets the reflection with respect to the line. A reflection with respect to a plane composed with a rotation of angle $\alpha = \pi$ determines the reflection with respect to the origin of coordinates.

Example 2.4.3 The transformation

$$\begin{pmatrix} 1 \\ x' \\ y' \\ z' \end{pmatrix} = \begin{pmatrix} 1 & 0 & 0 & 0 \\ b_1 & 1 & 0 & 0 \\ b_2 & 0 & 1 & 0 \\ b_3 & 0 & 0 & 1 \end{pmatrix} \begin{pmatrix} 1 \\ x \\ y \\ z \end{pmatrix}$$

represents the translation sending the origin to the point $P = (b_1, b_2, b_3)$.

Given a space curve parametrized by $(I, \alpha = (\alpha_1, \alpha_2, \alpha_3))$, the transformed curve admits $(I, \tilde{\alpha})$ as its parametrization, where

$$\tilde{\alpha}(t) = (\alpha_1(t) + b_1, \alpha_2(t) + b_2, \alpha_3(t) + b_3), \quad t \in I.$$

For example, the curve $((0, 2\pi), \alpha)$ given by

$$\alpha(t) = (\cos(2t)\sin(t), \cos(t), \sin(t))$$

is transformed into $((0, 2\pi), \tilde{\alpha})$, with $\tilde{\alpha}(t) = \alpha(t) + (1, 2, 3)$, after the translation of vector $(1, 2, 3)$.

The QR Code in Fig. 2.24 links to the translation of vector $v = (1, 2, 3)$ of the curve in Example 2.4.3.

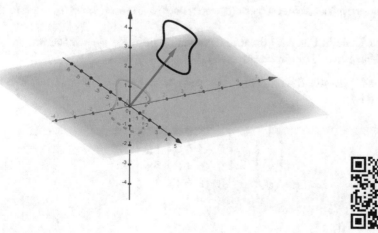

Fig. 2.24 Translation of vector $v = (1, 2, 3)$ of the curve in Example 2.4.3. QR Code 13

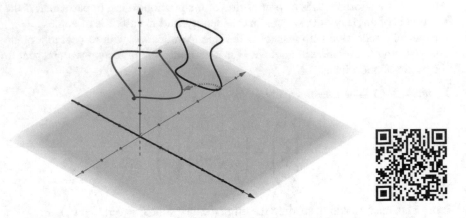

Fig. 2.25 Rotation of angle $\beta = \pi/4$ around $\{y = z = 0\}$ of the curve $(I, \tilde{\alpha})$ in Example 2.4.3.
QR Code 14

Example 2.4.4 The transformation

$$\begin{pmatrix} 1 \\ x' \\ y' \\ z' \end{pmatrix} = \begin{pmatrix} 1 & 0 & 0 & 0 \\ 0 & 1 & 0 & 0 \\ 0 & 0 & \cos(\beta) & -\sin(\beta) \\ 0 & 0 & \sin(\beta) & \cos(\beta) \end{pmatrix} \begin{pmatrix} 1 \\ x \\ y \\ z \end{pmatrix},$$

represents the rotation of angle β around the line $\{y = z = 0\}$.

In the curve of the previous example parametrized by $((0, 2\pi), \alpha)$ and for $\beta = \pi/4$, one gets the parametrization $((0, 2\pi), \alpha_s)$ with

$$\alpha_s(t) = (\cos(2t)\sin(t), \cos(\beta)\cos(t) - \sin(\beta)\sin(t), \sin(\beta)\cos(t) + \cos(\beta)\sin(t)), \quad t \in (0, 2\pi).$$

The QR Code in Fig. 2.25 links to the rotation of angle $\beta = \pi/4$ around $\{y = z = 0\}$ of the curve $(I, \tilde{\alpha})$ in Example 2.4.3.

Example 2.4.5 Lastly, the reflection with respect to the plane of equation $x = 0$ is determined by

$$\begin{pmatrix} 1 \\ x' \\ y' \\ z' \end{pmatrix} = \begin{pmatrix} 1 & 0 & 0 & 0 \\ 0 & -1 & 0 & 0 \\ 0 & 0 & 1 & 0 \\ 0 & 0 & 0 & 1 \end{pmatrix} \begin{pmatrix} 1 \\ x \\ y \\ z \end{pmatrix},$$

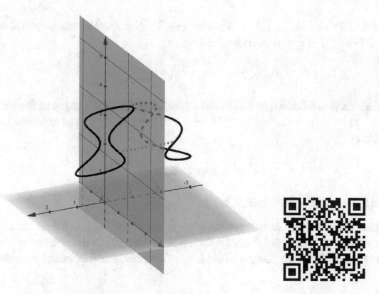

Fig. 2.26 Reflection with respect to the plane $x = 0$ of the curve $\tilde{\alpha}$ in Example 2.4.3. QR Code 15

The curve $((0, 2\pi), \alpha_s)$ in the previous example is transformed into $((0, 2\pi), \alpha_{sp})$ given by

$$\alpha_{sp}(t) = (-\cos(2t)\sin(t) - 1, \cos(t) + 2, \sin(t) + 3), \quad t \in (0, 2\pi).$$

The QR Code in Fig. 2.26 links to the reflection with respect to the plane $x = 0$ of the curve $\tilde{\alpha}$ in Example 2.4.3.

The next result describes how the coordinates of a point change when it is reflected with respect to a plane. In Sect. 4.4 (Theorem 4.4.1) we complete the information with the transformation of the coordinates of a point by performing a rotation around a line.

Proposition 2.4.6 *The reflection of a point (x, y, z) with respect to the plane of equation $ax + by + cz + d = 0$, with $(a, b, c) \neq (0, 0, 0)$, is determined by the coordinates (x', y', z'), with*

$$\begin{pmatrix} 1 \\ x' \\ y' \\ z' \end{pmatrix} = \Delta \begin{pmatrix} a^2 + b^2 + c^2 & 0 & 0 & 0 \\ -\frac{2ad}{\Delta} & -a^2 + b^2 + c^2 & -2ab & -2ac \\ -\frac{2bc}{\Delta} & -2ab & a^2 - b^2 + c^2 & -2bc \\ -\frac{2cd}{\Delta} & -2ac & -2bc & a^2 + b^2 - c^2 \end{pmatrix} \begin{pmatrix} 1 \\ x \\ y \\ z \end{pmatrix}$$

(2.30)

with $\Delta = (a^2 + b^2 + c^2)^{-1}$.

Proof Let $P = (x_0, y_0, z_0) \in \mathbb{R}^3$. The line at P normal to the plane of equation $ax + by + cz + d = 0$ is parametrized by

$$(x, y, z) = P + t(a, b, c), \quad t \in \mathbb{R}.$$

The intersection of that line and the plane takes place at the value of the parameter $t_0 = (-ax_0 - by_0 - cz_0 - d)/(a^2 + b^2 + c^2)$, and therefore the reflected point has coordinates

$$(x', y', z') = P + 2t_0(a, b, c),$$

which leads to the expression (2.30).

Concerning reflections and symmetries, in the framework of architectural studies, one can refer to Miltra and Pauly (2008), where the authors describe recent 3D tools applied in architectural designs to explore symmetry in architectural models.

2.5   Some Transformations on a Helix

In Sect. 3.4, we deal with modifications of helices to fit a surface. Here, we plan to observe and analyze the different elements involved in their parametrization and understand their nature in order to make modifications according to the necessities of a particular architectural element. This can be useful in order to understand many actions underlying the commands which can be performed in a CAD program.

We start from a regular parametrization of the circular helix $(\mathbb{R}, \alpha)$, with

$$\alpha(t) = (\cos(t), \sin(t), t), \quad t \in \mathbb{R}.$$

Its projections on the coordinate planes have been studied in Sect. 2.3, and displayed in Fig. 2.18. Its associated Frenet trihedron (see Definition 2.2.25) is given by

$$\left\{ \alpha(t); \left\{ \left(\frac{-\sin(t)}{\sqrt{2}}, \frac{\cos(t)}{\sqrt{2}}, \frac{1}{\sqrt{2}} \right), (-\cos(t), -\sin(t), 0), \left(\frac{\sin(t)}{\sqrt{2}}, \frac{-\cos(t)}{\sqrt{2}}, \frac{1}{\sqrt{2}} \right) \right\} \right\}. \tag{2.31}$$

The curvature and the torsion of the helix at any point are given by

$$\kappa_\alpha(t) = \frac{1}{2}, \quad \tau_\alpha(t) = \frac{1}{2}, \quad t \in \mathbb{R},$$

(see Propositions 2.2.21, 2.2.33 and the exercises of Chap. 2).

We first consider two modifications of the helix. The first one consists in drawing the helix, which grows in clockwise sense. This transformation is attained by performing a parametrization of the circle in plan clockwise, substituting t by $-t$

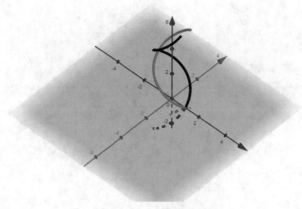

Fig. 2.27 Circular helix (black) vs. $(\mathbb{R}, \alpha_0)$ (green)

Fig. 2.28 Helix $(\mathbb{R}, \alpha_1)$ for $\rho = 1$ (black) vs. $\rho = 3$ (green)

in the two first coordinates of α. We have also taken into account the classical trigonometric properties to arrive at

$$\alpha_0(t) = (\cos(t), -\sin(t), t), \quad t \in \mathbb{R}$$

(see Fig. 2.27).

A second transformation is to consider the parametrization

$$\alpha_1(t) = (\rho \cos(t), \rho \sin(t), t), \quad t \in \mathbb{R}, \tag{2.32}$$

for some fixed $\rho > 0$. The plan view of this curve remains a circle centered at the origin, with radius ρ. The resulting helix grows in height at the same rate (see Fig. 2.28). In this case,

$$\kappa_{\alpha_1}(t) = \frac{\rho}{1 + \rho^2}, \quad \tau_{\alpha_1}(t) = \frac{1}{1 + \rho^2}, \quad t \in \mathbb{R}. \tag{2.33}$$

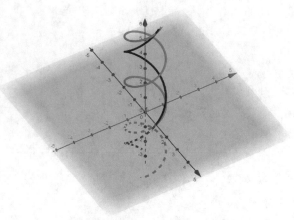

Fig. 2.29 Helix $(\mathbb{R}, \alpha_2)$ for $\rho = 1, h = 1$ (black) vs. $h = 1/2$ (green)

The growth rate of the helix is measured in the third coordinate. The modified helix parametrized by

$$\alpha_2(t) = (\rho \cos(t), \rho \sin(t), ht), \quad t \in \mathbb{R}, \tag{2.34}$$

for any $h > 0$ determines the velocity at which the helix grows in height. For example, a parametrization of a helix with $h = 1/2$ attains the point $(1, 0, 2\pi)$ at $t = 4\pi$, after two complete turns of the circle in plan view, whilst the initial helix (with $h = 1$) achieves that height after a single turn (see Fig. 2.29).

In this case, one has

$$\kappa_{\alpha_2}(t) = \frac{\rho}{h^2 + \rho^2}, \quad \tau_{\alpha_2}(t) = \frac{h}{h^2 + \rho^2}, \quad t \in \mathbb{R}. \tag{2.35}$$

The growth rate of the helix can be modified according to the requirements of a structure. In this regard, one can consider the curve parametrized by $(\mathbb{R}, \alpha_3)$, with

$$\alpha_3(t) = (\rho \cos(t), \rho \sin(t), h(t)), \quad t \in \mathbb{R}, \tag{2.36}$$

for some regular function $h : \mathbb{R} \to \mathbb{R}$. Let us fix $h(t) = t + \theta$, for some $\theta \in (0, 2\pi)$. The translation on the third variable causes a shifting on the height of the helix or equivalently a clockwise rotation of angle θ with respect to the line $\{x = y = 0\}$, when regarding the plane $z = 0$ from the semispace $z > 0$. Observe that

$$\alpha_3(t) = (\rho \cos(t), \rho \sin(t), t + \theta) = (\rho \cos(t + \theta - \theta), \rho \sin(t + \theta - \theta), t + \theta)$$

$$= (\rho \cos(\theta) \cos(t + \theta) + \rho \sin(\theta) \sin(t + \theta), \rho \cos(\theta) \sin(t + \theta) - \rho \sin(\theta) \cos(t + \theta),$$

$$t + \theta)$$

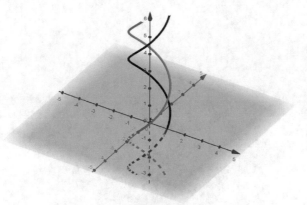

Fig. 2.30 Helix $(\mathbb{R}, \alpha_3)$ for $\rho = 1$, $h(t) = t$ (black) vs. $h(t) = t + \frac{\pi}{2}$ (green)

which leads to

$$\alpha_3(t)^T = \begin{pmatrix} 1 & 0 & 0 & 0 \\ 0 & \cos(\theta) & \sin(\theta) & 0 \\ 0 & -\sin(\theta) & \cos(\theta) & 0 \\ 0 & 0 & 0 & 1 \end{pmatrix} \alpha_1(t + \theta)^T.$$

Note that the previous matrix represents the rotation of angle $-\theta$ around the line $\{x = y = 0\}$ as shown in Example 2.4.4. Figure 2.30 shows both the initial helix and the transformed one, for $h(t) = t + \theta$, with $\theta = \frac{\pi}{2}$.

In the generic case, we obtain

$$\kappa_{\alpha_3}(t) = \frac{\rho\sqrt{(h''(t))^2 + (h'(t))^2 + \rho^2}}{(\rho^2 + (h'(t))^2)^{3/2}} \tag{2.37}$$

and

$$\tau_{\alpha_3}(t) = \frac{h'(t) + h'''(t)}{(h'(t))^2 + (h''(t))^2 + \rho^2}. \tag{2.38}$$

Other examples of these curves are given by $h(t) = t^3$ (Fig. 2.31), $h(t) = \exp(t)$ (Fig. 2.32) or $h(t) = \sin(t/3)$ (Fig. 2.33). Observe that the periodicity of the function in the last example gives rise to a bounded closed curve. The more general case $h(t) = \sin(At)$ describes a curve which turns out to be dense in a cilyndrical truncated surface, for $A \notin \mathbb{Q}$.

Let us consider the curve $(\mathbb{R}, \alpha_4)$ parametrized by

$$\alpha_4(t) = (\rho_1 \cos(t), \rho_2 \sin(t), t), \quad t \in \mathbb{R},$$

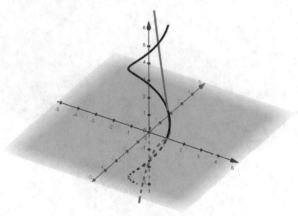

Fig. 2.31 Helix $(\mathbb{R}, \alpha_3)$ for $\rho = 1$, $h(t) = t$ (black) vs. $h(t) = t^3$ (green)

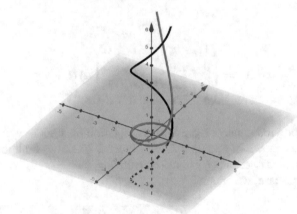

Fig. 2.32 Helix $(\mathbb{R}, \alpha_3)$ for $\rho = 1$, $h(t) = t$ (black) vs. $h(t) = \exp(t)$ (green)

for $\rho_1, \rho_2 > 0$. The parametrization of a conic, described in Sect. 1.4, shows that the plan view of that curve is an ellipse of semiaxes ρ_1 and ρ_2, respectively. The generalization can go further by choosing parametrizations of one of the branches of a hyperbola

$$\alpha_4(t) = (\rho_1 \cosh(t), \rho_2 \sinh(t), t), \quad t \in \mathbb{R},$$

or any other parametrizable plane curve. The shape of the curve obtained will no longer assemble a helix but the shape in plan view is the desired one. A final generalization would join the previous transformations to arrive at a parametrization of the form

$$\alpha_5(t) = (\rho_1(t) \cos(t), \rho_2(t) \sin(t), h(t)), \quad t \in \mathbb{R}, \tag{2.39}$$

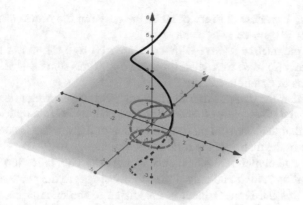

Fig. 2.33 Helix $(\mathbb{R}, \alpha_3)$ for $\rho = 1$, $h(t) = t$ (black) vs. $h(t) = \sin(t/3)$ (green)

allowing different distances to the line $\{x = 0, y = 0\}$ while moving in height with the function h.

This last approach is linked to cylindrical coordinates. It leads to different manners of designing more general space curves, described in Appendix A.

2.6 Suggested Exercises

2.1. Search for examples of space curves which cannot be embedded in a plane, constructed by adequate modifications of plane curves which are not regular curves, such as Example 2.1.5 from Example 1.1.10.

2.2. Find the geometry of the projections of the curve parametrized by $(\mathbb{R}, \alpha)$, with

$$\alpha(t) = \left(\cos(t), \sin(t), \frac{1}{2}\left(\exp\left(\sin(t)\right) + \exp\left(-\sin(t)\right)\right)\right), \quad t \in \mathbb{R}.$$

2.3. Find the curvature and torsion at any point in the loxodrome parametrized by (2.2). Draw the curve for some particular values of n, r and R and verify it is not a plane curve in that case. Does a loxodrome have inflection points?

2.4. Find the change of parameter leading to a natural parametrization of the curve parametrized by $\alpha(t) = (1, t, t^2)$, for $t \in \mathbb{R}$. Hint: Use an adequate hyperbolic change of variable in the corresponding integral.

2.5. Find necessary and sufficient conditions of $a_1, \ldots, a_9 \in \mathbb{R}$, with $a_2 \neq 0$, for the arc of curve parametrized by $\alpha : (0, \infty) \to \mathbb{R}^3$ with

$$\alpha(t) = (a_1 + a_2 t, a_3 + a_4 t + a_5 t^2, a_6 + a_7 t + a_8 t^2 + a_9 t^3)$$

that is a plane curve. Determine the plane in which the curve is contained in the positive situation.

2.6. Find the curvature at every point of the curve in the previous exercise, with
$a_2 = a_4 = a_5 = a_7 = a_8 = a_9 = 1$.

2.7. Find a parametrized curve with curvature $\kappa(t) \equiv c$ for all $t \in \mathbb{R}$ and torsion
given by $\tau(t) \equiv d$, for $t \in \mathbb{R}$, for some $c > 0$ and some $d \in \mathbb{R}$. Hint: Follow
the steps of Example 2.2.38.

2.8. Verify that the composition of rigid transformations which leave the origin
unchanged is associated to the product of their associated matrices.

2.9. Verify that the inverse rigid transformation of a given one is associated to the
inverse of the matrix of the rigid transformation.

2.10. Verify that a point remains unchanged after the reflection described in
Proposition 2.4.6 if and only if that point belongs to the plane of reflection.

2.11. Verify that the Frenet trihedron associated to the circular helix is given by
(2.31).

2.12. Verify that the curvature and torsion of the helix parametrized by (2.32) is
given by (2.33).

2.13. Do the same for (2.35) and (2.34).

2.14. Do the same for (2.37) together with (2.38) and (2.36). Verify that the
previous two exercises are consistent with the results obtained here.

2.15. Describe the geometry of a curve parametrized by (2.39), where $\rho_1(t) =
\rho_2(t) = h(t) = t$ for all $t \in \mathbb{R}$.

2.16. Find the curvature and torsion at every point of the curve in the previous
exercise.

2.17. Search for information about the following changes of coordinate systems:
parabolic cylindrical, paraboloidal, elliptic cylindrical and toroidal coordinate.

Chapter 3
Parametrizations and Regular Surfaces

After two chapters devoted to the study of curves, plane curves in Chap. 1, and space curves in Chap. 2, here we focus on the study of surfaces. This theory is also classic, and can be found with more detail in the book (do Carmo 1976), and explained with a more current approach in Tapp (2016), Umehara et al. (2017). Following the same structure as in the previous chapters, we will describe some transformations on surfaces which can be applied to or appear in architectural elements.

The symbolism, meaning or feeling that an architect is transmitting in an architectural work is captured in its shape, giving rise to particular forms which may correspond to or be inspired by classical mathematical surfaces. In this chapter and the following, we give several examples of these forms.

Again, the approaches to curves by means of parametric and implicit representations can be applied to surfaces. Each approach has benefits, depending on the final way to act on the surface itself, which resemble those described for curves. An implicit representation of a surface allows us to compute directly whether a point or other varieties belong to it. It also allows us to perturb the surface in order to obtain an approximated surface. In contrast, the parametric definition of a surface leads to a quick reconstruction of the surface by giving values to the parameters and provides local information on the variety.

The theory will be described in a way that resembles the corresponding theory related to curves, for the sake of coherence of the discourse.

3.1 Surfaces and Parametrizations

Intuitively, the notion of a surface can be associated to the deformation of (part of) a plane in space. One would say that the elements in Fig. 3.1 are surfaces.

© The Author(s), under exclusive license to Springer Nature Switzerland AG 2021
A. Lastra, *Parametric Geometry of Curves and Surfaces*, Mathematics and the Built
Environment 5, https://doi.org/10.1007/978-3-030-81317-8_3

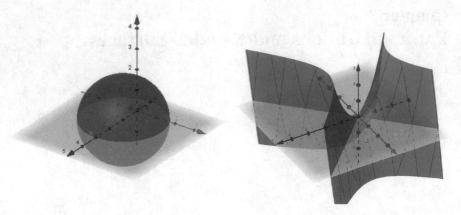

Fig. 3.1 Sphere centered at $(1, 0, 0)$ and radius $R = 2$ (left); hyperbolic paraboloid (right)

In the previous examples, the sphere is given by

$$S_1 = \{(x, y, z) \in \mathbb{R}^3 : (x - 1)^2 + y^2 + z^2 - 4 = 0\},$$

whereas the hyperbolic paraboloid corresponds to

$$S_2 = \{(x, y, z) \in \mathbb{R}^3 : x^2 - y^2 - z = 0\}.$$

These examples will lead to the implicit way to define a surface.

We may also consider a cone consisting of the set of all points $(x, y, z) \in \mathbb{R}^3$ such that

$$(x, y, z) = (v \cos(u), v \sin(u), v),$$

for some choice of $(u, v) \in \mathbb{R}^2$. A torus is determined by the points $(x, y, z) \in \mathbb{R}^3$ such that

$$(x, y, z) = (\cos(u)(3 + 2\cos(v)), \sin(u)(3 + 2\cos(v)), 2\sin(v)),$$

for some pair of values $(u, v) \in \mathbb{R}^2$. We have brought to light another way to define a surface by means of a parametrization.

Definition 3.1.1 A **parametrization** is a pair (U, X), where $\emptyset \neq U \subseteq \mathbb{R}^2$ is an open set of $\mathbb{R}^2$, and $X : U \to \mathbb{R}^3$ belongs to $\mathcal{C}^\infty(U)$.

We usually denote the components of X in the form

$$X(u, v) = (x(u, v), y(u, v), z(u, v)), \qquad (u, v) \in U. \tag{3.1}$$

We mainly work with the so-called regular parametrizations. This type of parametrization (U, X) determines sets in $\mathbb{R}^3$, $X(U)$, which are "essentially" open sets in $\mathbb{R}^2$.

Definition 3.1.2 A parametrization (U, X) is a **regular parametrization** if it holds that $X : U \to X(U)$ is a one-to-one mapping whose inverse $X^{-1} : X(U) \to U$ is a continuous function. Assuming (3.1), then for every $(u_0, v_0) \in U$, one has

$$\text{rank} \begin{pmatrix} \frac{\partial x}{\partial u}(u_0, v_0) & \frac{\partial y}{\partial u}(u_0, v_0) & \frac{\partial z}{\partial u}(u_0, v_0) \\ \frac{\partial x}{\partial v}(u_0, v_0) & \frac{\partial y}{\partial v}(u_0, v_0) & \frac{\partial z}{\partial v}(u_0, v_0) \end{pmatrix}^T = 2.$$

We recall that the first statement in the previous definition means that X^{-1} is the restriction of a continuous function $\tilde{X} : W \subseteq \mathbb{R}^3 \to \mathbb{R}^2$, for some open set $W \subseteq \mathbb{R}^3$ with $X(U) \subseteq W$. On the other hand, the last statement allows us to define the **tangent plane** to the surface parametrized by the regular parametrization at each of its points $X(u_0, v_0)$ because the vectors

$$\left(\frac{\partial x}{\partial u}(u_0, v_0), \frac{\partial y}{\partial u}(u_0, v_0), \frac{\partial z}{\partial u}(u_0, v_0) \right) \text{ and } \left(\frac{\partial x}{\partial v}(u_0, v_0), \frac{\partial y}{\partial v}(u_0, v_0), \frac{\partial z}{\partial v}(u_0, v_0) \right)$$

are linearly independent (see Fig. 3.2).

The previous condition is equivalent to the differential $dX_{(u_0,v_0)} : \mathbb{R}^2 \to \mathbb{R}^3$ being an injective mapping (see Marsden and Tromba (2012) for a deeper insight on this theory).

Our aim diverges from the deep topological aspects of the theory. However, it is worthwhile to mention that the word "essentially" when referring to $X(U)$ as a bent (part of a) plane corresponds to U and $X(U)$ being homeomorphic in the sense that there exists a continuous bijective map transforming one into the other. For the topological description of the previous concepts we refer to Munkres (1974).

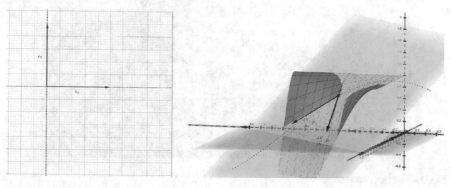

Fig. 3.2 $\left\{ \left(\frac{\partial x}{\partial u}(u_0, v_0), \frac{\partial y}{\partial u}(u_0, v_0), \frac{\partial z}{\partial u}(u_0, v_0) \right), \left(\frac{\partial x}{\partial v}(u_0, v_0), \frac{\partial y}{\partial v}(u_0, v_0), \frac{\partial z}{\partial v}(u_0, v_0) \right) \right\}$

Definition 3.1.3 Given a parametrization (U, X), we say that a point $X(u_0, v_0) \in$ $X(U)$ is a **singular point** if some of the local properties stated in Definition 3.1.2 are not satisfied. Otherwise, we say that $X(u_0, v_0)$ is a **regular point**.

Example 3.1.4 The points in the autointersections of the surface displayed in Fig. 3.3 cannot be regular points associated to a parametrization.

Definition 3.1.5 We say that $\emptyset \neq S \subseteq \mathbb{R}^3$ is a **regular surface** if for every $P \in S$ there exists a sphere centered at P, say S_P, and a regular parametrization (U, X) such that $X(U) = S_P \cap S$.

Consequently, a regular surface is one for which local coverings of the surface exist by means of the image of regular parametrizations (Fig. 3.4).

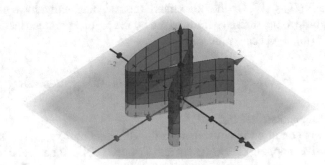

Fig. 3.3 Autointersection

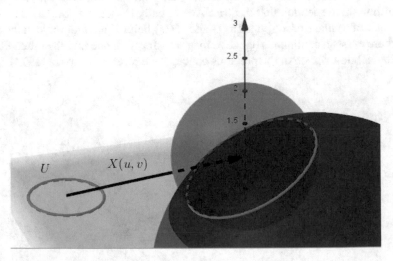

Fig. 3.4 Regular surface

Example 3.1.6 Let us consider the unit sphere. It is possible to cover all its points with the image of the parametrizations $(D((0, 0), 1), X_j)$ for $j = 1, \ldots, 6$, where

$$X_1(u, v) = (u, v, \sqrt{1 - u^2 - v^2}), \quad (u, v) \in D((0, 0), 1) \subseteq \mathbb{R}^2,$$

$$X_2(u, v) = (u, v, -\sqrt{1 - u^2 - v^2}), \quad (u, v) \in D((0, 0), 1) \subseteq \mathbb{R}^2,$$

$$X_3(u, v) = (u, \sqrt{1 - u^2 - v^2}, v), \quad (u, v) \in D((0, 0), 1) \subseteq \mathbb{R}^2,$$

$$X_4(u, v) = (u, -\sqrt{1 - u^2 - v^2}, v), \quad (u, v) \in D((0, 0), 1) \subseteq \mathbb{R}^2,$$

$$X_5(u, v) = (\sqrt{1 - u^2 - v^2}, u, v), \quad (u, v) \in D((0, 0), 1) \subseteq \mathbb{R}^2,$$

$$X_6(u, v) = (-\sqrt{1 - u^2 - v^2}, u, v), \quad (u, v) \in D((0, 0), 1) \subseteq \mathbb{R}^2.$$

The image of the first and second parametrizations leave the unit circle uncovered at the floor plan, which is totally covered by the third and fourth parametrizations, except for two points. These two points belong to the image of the fifth and sixth parametrizations. The different local coverings of the sphere are displayed in Fig. 3.5.

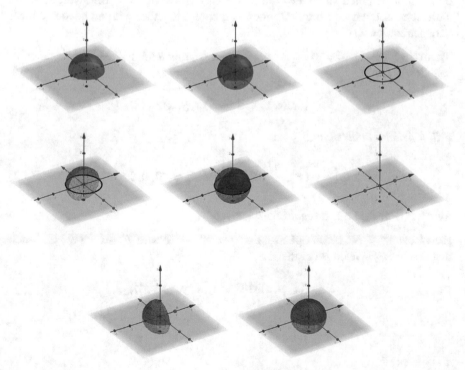

Fig. 3.5 Local coverings of the unit sphere

Fig. 3.6 Local coverings of the cone minus the vertex

In order to avoid singular points, we appeal to local parametrizations of surfaces.

Example 3.1.7 A cone is not a regular surface. The surface obtained by removing its vertex is a regular surface. Here are parametrizations covering the cone except for its vertex (Fig. 3.6).

$$X_1(u, v) = (v\cos(u), v\sin(u), v), \quad v > 0, u \in (0, 2\pi),$$

$$X_2(u, v) = (v\cos(u), v\sin(u), v), \quad v > 0, u \in (\pi, 3\pi),$$

$$X_3(u, v) = (v\cos(u), v\sin(u), -v), \quad v > 0, u \in (0, 2\pi),$$

$$X_4(u, v) = (v\cos(u), v\sin(u), -v), \quad v > 0, u \in (\pi, 3\pi).$$

Local parametrizations give rise to the definition of a regular surface described by an implicit function, as they did in the case of curves. The following result is a direct consequence of implicit function theorem together with an adequate local parametrization.

Theorem 3.1.8 *Let $\emptyset \neq W \subseteq \mathbb{R}^3$ be an open set, and let $F : W \to \mathbb{R}$ be a function in $\mathcal{C}^\infty(W)$. Let*

$$S = \{(x, y, z) \in \mathbb{R}^3 : F(x, y, z) = 0\}.$$

If $S \neq \emptyset$ and it holds that

$$\left(\frac{\partial F}{\partial x}(P), \frac{\partial F}{\partial y}(P), \frac{\partial F}{\partial z}(P)\right) \neq (0, 0, 0) \tag{3.2}$$

for all $P \in S$, then S is a regular surface.

Proof Let $W \subseteq \mathbb{R}^3$ be an open set and $F : W \to \mathbb{R}$ with $F \in \mathcal{C}^\infty(W)$ and such that $0 \in \text{Im}(F)$. We consider the set

$$S = \{(x, y, z) \in \mathbb{R}^3 : F(x, y, z) = 0\},$$

which is not empty. Let $P \in S$ and assume (3.2) holds. Then one may depart from $\frac{\partial F}{\partial z}(P) \neq 0$, without loss of generality. The implicit mapping theorem guarantees the existence of an open set $U \subseteq \mathbb{R}^2$ and $f : U \to \mathbb{R}$ with $f \in \mathcal{C}^\infty(U)$ such that

$$F(x, y, f(x, y)) = 0, \quad (x, y) \in U,$$

and such that there exists $(x_0, y_0) \in U$ with $f(x_0, y_0) = z_0$ and $P = (x_0, y_0, z_0)$.

Let $X_p : U \to \mathbb{R}^3$ be given by $X_p(u, v) = (u, v, f(u, v))$ for all $(u, v) \in U$. It holds that $X_p(U) \subseteq S$, $X_p \in \mathcal{C}^\infty(U)$, and $X_p^{-1} : X_p(U) \to \mathbb{R}^2$ is the projection with respect to its first two components, which is a continuous function on $X_p(U)$. In addition to this,

$$\text{rank} \begin{pmatrix} 1 & 0 & \frac{\partial f}{\partial u}(u, v) \\ 0 & 1 & \frac{\partial f}{\partial v}(u, v) \end{pmatrix}^T = 2, \quad (u, v) \in U.$$

This means that X_p is a local parametrization of S covering an open neighborhood of P in S.

In the proof, we have reduced the existence of local parametrizations to those that are the graph of a function in two variables (see Theorems 1.1.7 and 2.1.8 for analogous reasoning in the framework of plane and space curves). The parametrizations associated to a graph of a regular function are always regular. They are known as **Monge parametrizations**.

Proposition 3.1.9 *Let* $f : U \subseteq \mathbb{R}^2 \to \mathbb{R}^3$ *be a* $\mathcal{C}^\infty(U)$ *function defined on the open set* $\emptyset \neq U \subseteq \mathbb{R}^2$. *Then the* **graph** *of* f

$$S = \{(x, y, z) \in \mathbb{R}^3 : z - f(x, y) = 0\}$$

is a regular surface.

Proof The parametrization $X : U \to \mathbb{R}^3$ given by $(x, y) \mapsto (x, y, f(x, y))$ covers the whole graph of f, and it is a regular parametrization after analogous statements as those in the proof of the previous result.

In the study of curves we have the converse statements: Theorem 1.1.8 for plane curves and Theorem 2.1.9 when dealing with space curves. A local reciprocal result to Theorem 3.1.8 is also available in order to provide an alternative definition of a regular surface. Its proof follows a guideline analogous to that of Theorem 1.1.8, so we only provide an outline of it.

Theorem 3.1.10 *Let* S *be a regular surface. Then for every* $P = (x_0, y_0, z_0) \in S$ *there exists a sphere centered at* P *with positive radius, say* S_P, *and a scalar function* $F : S_P \to \mathbb{R}$ *with* $F \in \mathcal{C}^\infty(S_P)$ *such that*

$$S \cap S_P = \{(x, y, z) \in S_P : F(x, y, z) = 0\},$$

and

$$\nabla F(Q) = \left(\frac{\partial F}{\partial x}(Q), \frac{\partial F}{\partial y}(Q), \frac{\partial F}{\partial z}(Q) \right) \neq (0, 0, 0),$$

for every $Q \in S_P$.

Outline of the Proof Let $P = (x_0, y_0, z_0) \in S$, and choose a sphere $\tilde{S}_p \subseteq \mathbb{R}^3$ such that $\tilde{S}_p \cap S$ is described by a regular parametrization, say (U, X). Let us write

$$X(u, v) = (x(u, v), y(u, v), z(u, v)),$$

for all $(u, v) \in U$, and put $X(u_0, v_0) = P$ for some $(u_0, v_0) \in U$. From the rank condition on the partial derivatives of the components of a regular parametrization we can invert two of them, at least locally. Let us assume without loss of generality that it is the first two of them. We have the map $\varphi : U_1 \subseteq U \to \mathbb{R}^2$ defined on a non-empty open set U_1 of U by $\varphi(u, v) = (x(u, v), y(u, v))$ is invertible and $\varphi \in C^\infty(U_1)$, with $\varphi(u_0, v_0) = (x_0, y_0)$. Let $F : \varphi(U_1) \times (z_0 - \epsilon, z_0 + \epsilon) \to \mathbb{R}$ be defined in the following way

$$(x, y, z) \mapsto F(x, y, z) = z - z(\varphi^{-1}(x, y)),$$

for some $\epsilon > 0$.

The above result leads to an alternative definition of a regular surface:

Definition 3.1.11 Let $\tilde{S} \subseteq \mathbb{R}^3$ be a nonempty open set. Let $F : \tilde{S} \to \mathbb{R}$ with $F \in C^\infty(U)$. We say that the set

$$S = \{(x, y, z) \in \tilde{S} : F(x, y, z) = 0\}$$

is a **regular surface** if $S \neq \emptyset$ and for every $P \in S$ it holds that

$$\nabla F(P) = \left(\frac{\partial F}{\partial x}(P), \frac{\partial F}{\partial y}(P), \frac{\partial F}{\partial z}(P) \right) \neq (0, 0, 0).$$

As in the case of curves, it does not hold that the image of any regular parametrization describes a regular surface. The condition on the existence of a sphere for each point (see Definition 3.1.5) is crucial in this sense. See Example 1.1.10 in the framework of plane curves.

A reparametrization of $X(U)$ associated to a regular parametrization (U, X) in this context turns out to be an invertible mapping $\varphi : U \to V \subseteq \mathbb{R}^2$, with $\varphi \in C^\infty(U)$. The pair $(V, X \circ \varphi^{-1})$ is a regular parametrization of $X(U)$. It can also be proved that all reparametrizations are of this form.

In the context of plane and space curves, we have detailed a way to compute the arc length between two points of the curve. The notion of curvature and torsion were first given for arc length parametrizations. Some elements related to the surface such as areas, distances or angles can also be computed here. More precisely, the **first fundamental form** associated to a regular surface acts as the scalar product restricted to the surface. Therefore, it allows us to compute angles and distances within the surface.

Let (I, α) be a regular curve contained in a regular surface parametrized by (U, X), such that $\alpha(t) = X(u(t), v(t))$ for $t \in I$ under adequate compatibility conditions on the definition of the previous composition. Then it holds that the arc length of $\alpha(I)$ is given by

$$\int_I \|\alpha'(t)\| \, dt$$

(see Proposition 2.2.11). Now observe that

$$\|\alpha'(t)\|^2 = \alpha'(t) \cdot \alpha'(t)$$

$$= \left(\frac{\partial \alpha}{\partial u}(u(t), v(t))u'(t) + \frac{\partial \alpha}{\partial v}(u(t), v(t))v'(t) \right) \cdot$$

$$\left(\frac{\partial \alpha}{\partial u}(u(t), v(t))u'(t) + \frac{\partial \alpha}{\partial v}(u(t), v(t))v'(t) \right)$$

$$= E(u'(t))^2 + 2Fu'(t)v'(t) + G(v'(t))^2 = \left(u'(t) \; v'(t) \right) \begin{pmatrix} E & F \\ F & G \end{pmatrix} \begin{pmatrix} u'(t) \\ v'(t) \end{pmatrix},$$

where $E = \frac{\partial \alpha}{\partial u}(u(t), v(t)) \cdot \frac{\partial \alpha}{\partial u}(u(t), v(t))$, $F = \frac{\partial \alpha}{\partial u}(u(t), v(t)) \cdot \frac{\partial \alpha}{\partial v}(u(t), v(t))$, $G = \frac{\partial \alpha}{\partial v}(u(t), v(t)) \cdot \frac{\partial \alpha}{\partial v}(u(t), v(t))$. The matrix

$$\begin{pmatrix} E & F \\ F & G \end{pmatrix}$$

represents the inner product restricted to the surface, i.e., the first fundamental form of the surface.

It is also possible, following arguments similar to those applied in Proposition 1.3.14 to derive a formula to measure areas in a regular surface. More precisely, given the regular parametrization (U, X), and a nonempty open set $V \subseteq U$, then the area of $X(V)$ is determined by

$$\int \int_V \left| \frac{\partial X}{\partial u}(u, v) \times \frac{\partial X}{\partial v}(u, v) \right| du dv, \tag{3.3}$$

which can also be written in terms of the first fundamental form. It can be proved by change of coordinates that (3.3) does not depend on the regular parametrization considered.

In order to introduce the concepts of the normal vector and the tangent plane to a regular surface at a point we first state the concept of **arc of curve contained a surface**.

Definition 3.1.12 We say the space curve C determined by the parametrization (I, α) is an arc of curve contained in a regular surface S if $\alpha(I) \subseteq S$.

Example 3.1.13 Let us consider the cylinder

$$S = \{(x, y, z) \in \mathbb{R}^3 : x^2 + y^2 = 1\},$$

and the helix associated to the regular parametrization $(\mathbb{R}, \alpha)$, with

$$\alpha(t) = (\cos(t), \sin(t), t), \quad t \in \mathbb{R}.$$

It is clear that $\alpha(\mathbb{R}) \subseteq S$ so the helix is an arc of regular curve contained in the cylinder.

Observe moreover that any point in the cylinder can be parametrized by

$$X(u, v) = (\cos(u), \sin(u), v), (u, v) \in \mathbb{R}^2,$$

which yields that $\alpha(t) = X(t, t)$ for every $t \in \mathbb{R}$. See Fig. 3.7.

The two next results are of great importance. We have decided to include an outline of the proof, while the details can be found in Costa et al. (1997). The first result allows us to locate regular space curves contained in a regular surface from parametrizations of plane curves whose image is contained in the domain of definition of the parametrization of the surface. The second result will help when the time comes to give a definition of the tangent plane to a surface at a point.

Proposition 3.1.14 *Let (U, X) be a regular parametrization of a regular surface S, and let (I, α) be a regular parametrization of a plane curve, with $\alpha(I) \subseteq U$. Then $(I, X \circ \alpha)$ is a regular parametrization of an arc of regular curve contained in S.*

Fig. 3.7 Helix contained in a cylinder in Example 3.1.13

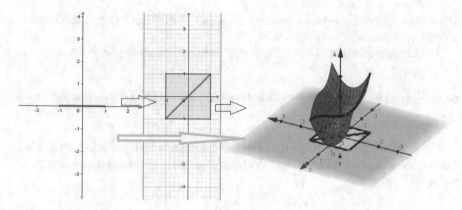

Fig. 3.8 Scheme of the construction of the regular curve contained in a regular surface

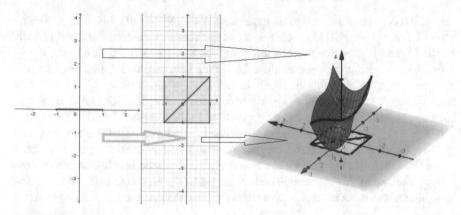

Fig. 3.9 Scheme of the construction of a regular plane curve from a regular curve contained in a surface

Outline of the Proof The proof consists in constructing the space curve contained in the surface S by means of composition $X \circ \alpha$ (See Fig. 3.8), which turns out to describe a regular space curve whose derivative at any value of the parameter does not vanish.

Proposition 3.1.15 *Let* (I, α) *be a regular parametrization of a space curve contained in a given regular surface S, associated to the regular parametrization* (U, X). *Then for all* $P = \alpha(t_0) \in S$, *there exists an interval* $J \subseteq I$, *with* $t_0 \in J$ *such that* $(J, X^{-1} \circ \alpha)$ *is a regular parametrization of a plane curve contained in* U.

Outline of the Proof Local invertibility of X at P is guaranteed from the regularity of the surface. The adequate reduction of the interval I allows us to compute $X^{-1} \circ \alpha$, which is a regular parametrization of a plane curve inside U. Figure 3.9 illustrates the procedure.

Definition 3.1.16 Let S be a regular surface, and $P \in S$. Let $\Gamma(P; S)$ be the set of all regular curves at P contained in S.

The **tangent plane** to the surface S at P is the affine variety at P associated to the subspace

$$\{\alpha'(t_0) : (I, \alpha) \text{ is a regular parametrization of an element in } \Gamma(P; S), \quad \alpha(t_0) = P\}.$$

We denote it by $T_S(P)$.

There are some points about the previous definition which are worth remarking. First, the vector space associated to the tangent plane is always of dimension 2 and is determined by

$$L\left(\left\{\frac{\partial X}{\partial u}(u_0, v_0), \frac{\partial X}{\partial v}(u_0, v_0)\right\}\right), \qquad (3.4)$$

where (U, X) stands for a regular parametrization of (part of) the surface S, with $P = X(u_0, v_0) \in X(U)$, $(u, v) \mapsto X(u, v)$. This last statement can be directly verified when it is realized that the curves $(u, v_0) \mapsto X(u, v_0)$ and $(u_0, v) \mapsto X(u_0, v)$ are regular curves contained in S (see Proposition 3.1.14), with tangent vectors being the elements in (3.4). In addition, given a curve (I, α) which is contained in S and such that $\alpha(t_0) = P$ for some $t_0 \in I$, it holds that $\alpha = X \circ \gamma$, for some regular plane curve (J, γ), in view of Proposition 3.1.15. This means that $\alpha'(t_0) = dX_{(u_0, v_0)}(\gamma'(t_0))$ and therefore $\alpha'(t_0)$ is a linear combination of the vectors in (3.4). Figure 3.10 illustrates a tangent plane to a surface at some specific point.

At this point, it is worthwhile to say something about the first fundamental form, already defined in this chapter. When restricted to the elements in $T_S(P)$, the first fundamental form can be applied to such vectors as an inner product. More precisely,

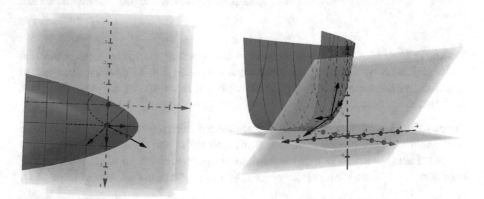

Fig. 3.10 Tangent plane to $z - \exp((x - 1)^2 + y) = 0$ at $P = (1, 0, 1)$

given a vector $v = (v_1, v_2, v_3) \in T_S(P)$, then there exist $x_1, x_2 \in \mathbb{R}$ such that

$$(v_1, v_2, v_3) = x_1 \frac{\partial X}{\partial u}(u_0, v_0) + x_2 \frac{\partial X}{\partial v}(u_0, v_0).$$

The first fundamental form of v is defined by the number

$$I(v) = \begin{pmatrix} x_1 & x_2 \end{pmatrix} \begin{pmatrix} E & F \\ F & G \end{pmatrix} \begin{pmatrix} x_1 \\ x_2 \end{pmatrix}. \tag{3.5}$$

We have also achieved the next result.

Corollary 3.1.17 *Let S be a regular surface parametrized by (U, X). Then for all $P = X(u_0, v_0) \in S$ the tangent plane to S at $P = X(u_0, v_0)$ is defined by*

$$\begin{vmatrix} x - x(u_0, v_0) & y - y(u_0, v_0) & z - z(u_0, v_0) \\ \frac{\partial x}{\partial u}(u_0, v_0) & \frac{\partial y}{\partial u}(u_0, v_0) & \frac{\partial z}{\partial u}(u_0, v_0) \\ \frac{\partial x}{\partial v}(u_0, v_0) & \frac{\partial y}{\partial v}(u_0, v_0) & \frac{\partial z}{\partial v}(u_0, v_0) \end{vmatrix} = 0,$$

with $X(u, v) = (x(u, v), y(u, v), z(u, v))$, for all $(u, v) \in U$.

Definition 3.1.18 Let S be a regular surface and $P \in S$. Either one of the two unit vectors which are orthogonal to the tangent plane of S at P is known as the **normal vector** to the surface S at P.

In many references, it is usual to distinguish one of these two vectors in order to give an orientation to the surface. This concept is very important in applications. We refer to the study of Möbius band and bridges in Sect. 3.2 for a brief summary of this topic.

The next result is a consequence of the definition of the cross product of two vectors.

Corollary 3.1.19 *Given a regular surface S and $P \in S$, the vector*

$$\frac{\frac{\partial X}{\partial u}(u_0, v_0) \times \frac{\partial X}{\partial v}(u_0, v_0)}{\left\| \frac{\partial X}{\partial u}(u_0, v_0) \times \frac{\partial X}{\partial v}(u_0, v_0) \right\|} \tag{3.6}$$

is a normal vector to S at $P = X(u_0, v_0)$.

The concept of normal vector to a regular surface can be stated when starting from an implicit definition of the surface (Fig. 3.11).

Proposition 3.1.20 *Let*

$$S = \{(x, y, z) \in \mathbb{R}^3 : F(x, y, z) = 0\}$$

be a regular surface. A normal vector to S at $P \in S$ is given by

$$\frac{\left(\frac{\partial F}{\partial x}(P), \frac{\partial F}{\partial y}(P), \frac{\partial F}{\partial z}(P) \right)}{\left\| \left(\frac{\partial F}{\partial x}(P), \frac{\partial F}{\partial y}(P), \frac{\partial F}{\partial z}(P) \right) \right\|},$$

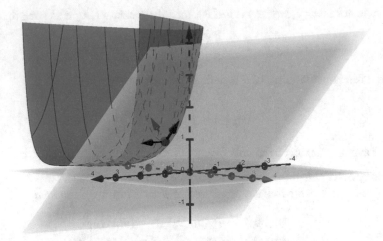

Fig. 3.11 Normal vector to $X(u, v) = (u, v, e^{(u-1)^2+v})$ at $P = (1, 0, 1)$

and the tangent plane to S at $P = (x_0, y_0, z_0) \in S$ is given by

$$\frac{\partial F}{\partial x}(P)(x - x_0) + \frac{\partial F}{\partial y}(P)(y - y_0) + \frac{\partial F}{\partial z}(P)(z - z_0) = 0.$$

Proof Let $P \in S$ and let (U, X) be a regular parametrization of S, with $X(u_0, v_0) = P$. We write $X(u, v) = (x(u, v), y(u, v), z(u, v))$, as usual. Then the vector

$$\left(\frac{\partial F}{\partial x}(P), \frac{\partial F}{\partial y}(P), \frac{\partial F}{\partial z}(P) \right)$$

is orthogonal to both $\frac{\partial X}{\partial u}(u_0, v_0)$ and $\frac{\partial X}{\partial v}(u_0, v_0)$. This last claim is easily verified by taking derivatives at the identity $F(x(u, v), y(u, v), z(u, v)) = 0$, valid for all $(u, v) \in U$, with respect to u and with respect to v.

3.2   Some Classic Surfaces in Architecture

In this section, we analyze some classic surfaces appearing in architectural elements.

Toroid Structures, and le Comptoir forestier
In Samyn (2017, pp.69–70), available in Samyn & Partner's webpage, Philippe Samyn describes the canopy of Comptoir forestier, a building formed by a torus section located in the region of Marche-en-Famenne, Belgium, and designed by Samyn & Partners. This is an example of a classic surface inspiring an architectural design. Several nice photographs of this building can also be obtained from their webpage (Samynandpartners 2020b).

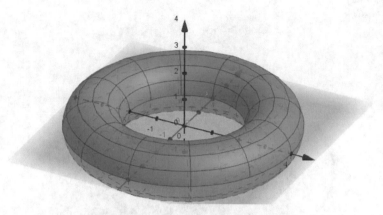

Fig. 3.12 Torus

The mathematical model lying behind this building is the surface known as a **torus**. This surface, represented in Fig. 3.12, can be generated by moving a circle (of radius $r > 0$) whose center is a point of a second circle (of radius $R > r$). The planes containing each circle are orthogonal. This situation can be parametrized as follows. Let R be the radius of the fixed circle, and assume that the origin of coordinates is the center of that circle. Moreover, we assume that it is contained in the floor plane. Any point of that circle is given by

$$(x, y, z) = (R\cos(\varphi), R\sin(\varphi), 0), \quad \varphi \in \mathbb{R}. \tag{3.7}$$

Let $\varphi_0 \in \mathbb{R}$ be fixed.

We now search for the plane which is orthogonal to the floor plane, passing through the origin of coordinates and the point $(R\cos(\varphi_0), R\sin(\varphi_0), 0)$. That plane is associated to an equation of the form $ax + by + cz - d = 0$, for some $a, b, c, d \in \mathbb{R}$. However, $d = 0$ because the origin of coordinates belongs to the plane. In addition to this, the normal vector of the floor plane is $(0, 0, 1)$, which is orthogonal to the vector (a, b, c). This means $c = 0$. In order for the point $(R\cos(\varphi_0), R\sin(\varphi_0), 0)$ to belong to the plane, we find that $a = R\sin(\varphi_0)$ and $b = -R\cos(\varphi_0)$ are valid choices for the parameters. Let us consider the sphere of radius $r > 0$, centered at $(R\cos(\varphi_0), R\sin(\varphi_0), 0)$. The intersection of the previous plane and the sphere describes the position of the moving circle at the value of the parameter $\varphi = \varphi_0$ (Fig. 3.13). In order to obtain its equation, we solve the system determined by both surfaces

$$\begin{cases} \sin(\varphi_0)x - \cos(\varphi_0)y = 0 \\ (x - R\cos(\varphi_0))^2 + (y - R\sin(\varphi_0))^2 + z^2 = r^2 \end{cases} \tag{3.8}$$

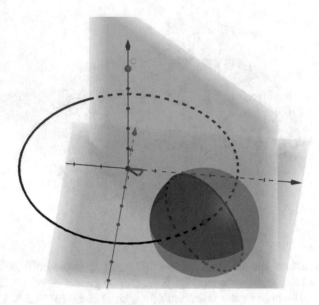

Fig. 3.13 Geometric construction of the torus

The second equation can be expanded to get

$$x^2 + y^2 + z^2 + R^2 - r^2 = 2R(\cos(\varphi_0)x + \sin(\varphi_0)y).$$

Taking squares of the previous equation yields

$$(x^2 + y^2 + z^2 + R^2 - r^2)^2 = 4R^2(\cos(\varphi_0)x + \sin(\varphi_0)y)^2.$$

Now, the first equation in (3.8) is equivalent to the fact that

$$(x, y) \perp (- \sin(\varphi_0), \cos(\varphi_0)),$$

so (x, y) and $(\cos(\varphi_0), \sin(\varphi_0))$ are linearly dependent, which means that

$$\cos(\varphi_0)x + \sin(\varphi_0)y = (x, y) \cdot (\cos(\varphi_0), \sin(\varphi_0)) = \pm\sqrt{x^2 + y^2}.$$

From that, we can conclude that an implicit equation defining the torus is

$$S = \{(x, y, z) \in \mathbb{R}^3 : (x^2 + y^2 + z^2 + R^2 - r^2)^2 = 4R^2(x^2 + y^2)\}.$$

The parametric equations of the torus can also be derived from the geometrical situation described, and follow the construction of the spherical coordinate system (see Appendix A). More precisely, any point in the torus can be determined as follows: any point in the fixed circle is given by (3.7). Now, from this position we can move to all the points in the moving circle, which in spherical coordinates is given by

$$(x, y, z) = (r\cos(\theta)\cos(\varphi), r\cos(\theta)\sin(\varphi), r\sin(\theta)), \quad \theta \in \mathbb{R},$$

with the point in (3.7) being the origin of the spherical coordinate system. We derive that

$$\begin{cases} x = R\cos(\varphi) + r\cos(\theta)\cos(\varphi) \\ y = R\sin(\varphi) + r\cos(\theta)\sin(\varphi) \\ z = \qquad\qquad r\sin(\theta) \end{cases} \tag{3.9}$$

is a parametrization of the torus, except for two circles.

Observe that the surface is completely reconstructed for every interval of length 2π for each of the parameters.

The implicit equation of a torus can also be obtained from the parametric one by means of implicitation results such as those in Lastra et al. (2018), as described previously.

Other buildings inspired in toroidal surfaces are Dubai's Museum of the Future (maybe it is finished when reading this book!), by Shaun Killa of Killa Design (see Fig. 3.14); a water tower in Ciechanow, Poland, by architect Jerzy Michał Bogusławski or Phoenix Media Center in Beijing, by Shau Weiping of BIAD.

Fig. 3.14 Dubai's Museum of the Future by Killa Design

Fig. 3.15 QR Code 16

Möbius Band, and Bridges

So far in this chapter, we have not made any distinction between the two possible normal vectors associated to a regular surface at a point. A concept related to the possibility of a choice of normal vectors at a point in some adecquate manner is the **orientability** of the surface. Generaly speaking, an orientable surface is one in which two sides can be distinguished, which is related to a subfamily of normal vectors.

A sphere and a torus are examples of orientable surfaces. However, a **Möbius band** is not. This surface can be constructed as follows. Let C be a circle and I a finite segment whose middle point belongs to the circle and remains orthogonal to the circle at that point. Now, move the segment along the circle, by preserving the point of contact while rotating the segment, preserving orthogonality with respect to the circle on this movement. The rotation rate is such that at the moment of completing the first loop, the segment arrives at the initial point after a half turn. The points described by the segment in this movement give rise to the Möbius band. We refer to the QR Code of Fig. 3.15.

We proceed to describe a parametrization of such surface. Assume the circle is centered at the origin of coordinates, is of radius $R > 0$, and is located at the floor plane. The segment, of length $2r > 0$, is initially touching the circle at the point $(R, 0, 0)$, and is also in horizontal position. The contact point moves with the parametrization of the circle

$$(x, y, z) = (R\cos(\theta), R\sin(\theta), 0).$$

Observe that the slope of the segment at a particular value of θ is $\theta/2$ in order to perform half a turn after the contact point arrives at the origin point with the movement. The relative position of the segment at θ is described in Fig. 3.16. Therefore, any point of the segment has the following coordinates with respect to the contact point:

$$(x, y, z) = (u\cos(\theta/2)\cos(\theta), u\cos(\theta/2)\sin(\theta), u\sin(\theta/2)), \quad -r < u < r.$$

Therefore, the parametrization of the Möbius band is given by

$$\begin{cases} x = R\cos(\theta) + u\cos(\frac{\theta}{2})\cos(\theta) \\ y = R\sin(\theta) + u\cos(\frac{\theta}{2})\sin(\theta) \\ z = \qquad\qquad u\sin(\frac{\theta}{2}), \qquad u \in \mathbb{R}, u \in (-r, r) \end{cases} \tag{3.10}$$

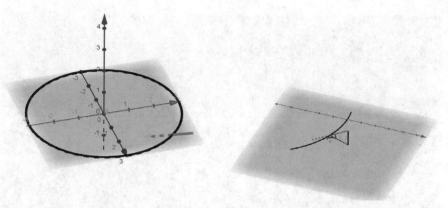

Fig. 3.16 Detail of the construction of a Möbius band

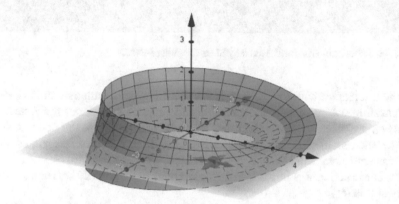

Fig. 3.17 Möbius band

Observe that the construction of the surface is repeated for every interval of length 2π in the parameter u. An implicitation procedure (see Lastra et al. (2018) or Sendra et al. (2007) for an in-depth approach) yields the implicit equation describing the Möbius band:

$$S = \{(x, y, z) \in \mathbb{R}^3 : -R^2 y + x^2 y + y^3 - 2Rxz - 2x^2 z - 2y^2 z + yz^2 = 0\}.$$

Note the previous surface describes the complete Möbius band for an infinite segment I (Fig. 3.17).

At this point, we can provide a more comprehensive manner to explain that the Möbius band is not an orientable surface. Assume that one considers one of the normal vectors associated to the surface at the point $(R, 0, 0)$. The continuous motion of the normal vector with the segment within the Möbius band leads to the opposite normal vector when a loop around the fixed circle has been completed. Therefore, the orientation has changed after traversing the surface.

Fig. 3.18 Phoenix International Media by Shau Weiping of BIAD

The fascination with this surface is usually due to its aesthetic qualities, due to the puzzling nature of this surface. The same can be said about the influence it exerts on architectural elements. It has appeared recursively in the construction of bridges such as the Möbius Bridge in Bristol, UK, BY Hakes Associates, or the more recent Lucky Knot Bridge, in Changsha, China, designed by Next Architects, whose construction was finished in 2016. We refer to Séquin (2018) for an in-depth study of this topic.

The building Phoenix International Media in Beijing, is deeply inspired by the Möbius band. This building was designed by Beijing Institute of Architectural Design (BIAD) and Un-Forbidden office, and was completed in 2013 (Fig. 3.18). We also refer to Thulaseedas and Krawczyk (2003) for different concepts related to this geometry in architecture.

Other manifestations of this surface appear in sculptures such as *Unendliche Schleife* by Max Bill, in the Centre Pompidou, Paris. In Frazier and Schattschneider (2008), Larry Frazier and Doris Schattschneider describe realizations of this surface in wood and alabaster. Möbius bridges and buildings are also mentioned in Séquin (2008).

Another non-orientable surface is Klein bottle. This surface can be parametrized (see Gray (1997)) by

$$\begin{cases} x = (r + \cos(\theta/2)\sin(v) - \sin(\theta/2)\sin(2v))\cos(\theta) \\ y = (r + \cos(\theta/2)\sin(v) - \sin(\theta/2)\sin(2v))\sin(\theta) \\ z = \sin(\theta/2))\sin(v) + \cos(\theta/2)\sin(2v) \end{cases}$$

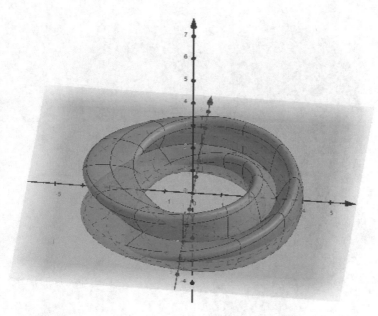

Fig. 3.19   Klein bottle. $r = 3$

for fixed $r > 2$, and $0 < \theta, v < 2\pi$. See Fig. 3.19. We thank Kim Williams for making us aware of the Klein Bottle House by Robert McBride and Debbie Ryan, in Australia, that can be found in Wikiarquitectura (2021a).

3.3   Projections of Surfaces onto Planes

The theory behind this section is broad enough to cover the whole book, and we have decided to focus on the immediate applications to architecture as a tool in order to determine the parametrization of a surface or a space curve from their different views.

Several representation systems have been considered while drawing 3D figures in a sheet of paper. This procedure is based on the choice of a projection of the object onto a plane. Depending on the relative position of the plane with respect to the projection one can distinguish between orthographic or oblique projections. The first of these deals with projections in which the projecting plane is orthogonal with respect to the direction of projection. Representation systems are usually orthographic: it suffices to fix the representation of the origin of coordinates in the projection plane and the projection of the axis OX, OY and OZ, together with a scale related to its axis. There are certain conditions for a projection to be orthonormal regarding the previous choices, and in this direction of great interest is the so-called Gauss fundamental theorem of normal axonometry (see Gauss

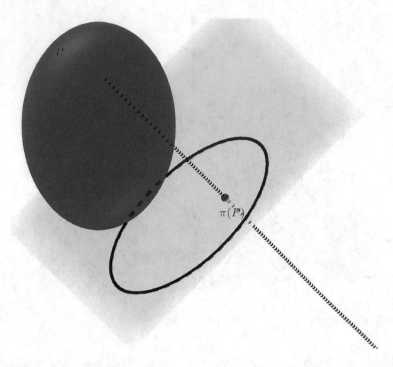

Fig. 3.20 Orthogonal projection π on a plane

(1876), and also Eastwood and Penrose (2000), which describes this result in higher dimensions). A direct consequence of this result is the Weisbach theorem (also known as Schlömilch Theorem, see Dörrie (2013)) which states rules for admissible scales at each axis. Let m_1, m_2, m_3 be the scaling factors associated to the representation of each axis. Isometric projection states $m_1 = m_2 = m_3 = \sqrt{2/3}$, with angles between each pair of representation of the axis of 120°. Dimetric projection deals with two equal scales and angles, whilst trimetric projection considers all scales to be different from each other.

Although isometric projection preserves the same scale with respect to each projection of the coordinate axis, it might be of interest to distinguish two of the views or other appropriate properties of a single projection (Fig. 3.20).

A deep insight into the construction of different projection systems from the approach of linear algebra can be followed in the classic work by Felix Klein (2004).

For practical reasons, we are only considering orthogonal projections of the surfaces and curves onto the coordinate planes $\{x = 0\}$, $\{y = 0\}$ and $\{z = 0\}$. However, the techniques used could also be effective for any other choice of the planes, depending on the specific necessities.

Let us consider the National Center for the Performing Arts, in Beijing. This building is an ellipsoid dome made of titanium and glass, by the architect Paul Andreu (Fig. 3.21).

Fig. 3.21 National Center for the Performing Arts in Beijing by Paul Andreu

Starting from the view of the arts center in Fig. 3.21, and assuming that the picture was taken with a camera whose lens was parallel to the floor, and by means of the approximation of the floor plan obtained from a different picture, we can approximate the view in Fig. 3.21 with the ellipse of equation

$$a_{11}x^2 + 2a_{12}xz - a_{22}z^2 + 2a_{02}z + a_{00} = 0, \tag{3.11}$$

with

$$a_{11} = -5979.2482935398, \quad 2a_{12} = 299.8749891932,$$
$$a_{22} = 29702.7281472082,$$
$$2a_{02} = 8224.3496907795, \quad a_{00} = 400085.853516655.$$

Observe in Fig. 3.22 that the approximation seems quite adequate, naturally dismissing the reflection of the building in the water. The same procedure gives rise to the equation of the approximated ellipse in the floor plane:

$$b_{11}x^2 + 2b_{12}xy + b_{22}y^2 + b_{00} = 0, \tag{3.12}$$

with

$$b_{11} = 40.4338103204, \quad 2b_{12} = -0.2839525946,$$
$$b_{22} = 82.6911856839, \quad b_{00} = -2705.5232896795.$$

It holds that if such ellipses provide the projections onto two of the coordinate planes of the ellipsoid determining the arts center, then the coefficients a_{11} and b_{11}

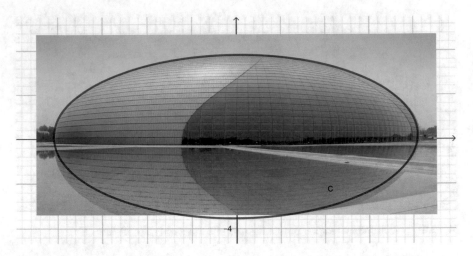

Fig. 3.22 Approximation of the National Centre for the Performing Arts, in Beijing. Photograph transformed from the previous

should coincide. For this reason, we multiply the Eq. (3.12) by a_{11}/b_{11}. The constant terms in the Eq. (3.11), and $b_{00}a_{11}/b_{11}$ should also coincide. We observe that

$$\left| a_{00} - \frac{b_{00}a_{11}}{b_{11}} 1 \right| \approx 0.0000004806.$$

This small error is caused by measurements made in the projections. Taking into account (3.11) and (3.12) one can reconstruct the equation of the ellipsoid by joining the terms in both equations:

$$a_{11}x^2 + 2b_{12}\frac{a_{11}}{b_{11}}xy + b_{22}\frac{a_{11}}{b_{11}}y^2 + 2a_{12}xz - a_{22}z^2 + 2a_{02}z + a_{00} = 0 \quad (3.13)$$

which represents the arts center under study.

We observe that the projection of the ellipsoid onto the coordinate plane $z = 0$ consists in evaluating (3.13) at $z = 0$, recovering the ellipse (3.12), whereas the evaluation $y = 0$ describes the ellipse in (3.11).

Figure 3.23 illustrates the result.

3.4 Curves in Surfaces and Intersection of Surfaces

This section is devoted to briefly describing curves which are part of a given surface, and some of their applications in architecture. Afterwards, we will also say a few words about the intersection of regular surfaces.

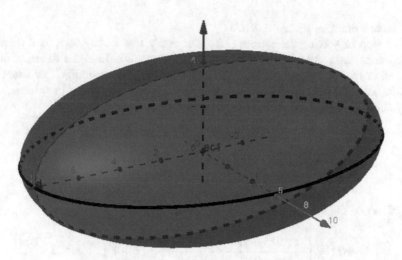

Fig. 3.23 Ellipsoid approximating of the National Centre for the Performing Arts, in Beijing

In this book, we have mainly provided curves and surfaces in two different ways, that is by means of their parametrization or by means of implicit functions. As a matter of fact, if a regular parametrization of a curve (I, α) is given, with $\alpha = (\alpha_1(t), \alpha_2(t), \alpha_3(t))$, and a regular parametrization of the surface (U, X), with $X = (x(u, v), y(u, v), z(u, v))$, is also provided then the curve is completely contained in the surface if it holds that for every $t \in I$, there exists $(u, v) \in U$ such that

$$\begin{cases} x(u, v) = \alpha_1(t) \\ y(u, v) = \alpha_2(t) \\ z(u, v) = \alpha_3(t). \end{cases} \tag{3.14}$$

Taking into account that no autointersections are allowed in the surface, given $t \in I$, the solution of the system in (3.14) in (u, v), if it exists, is unique. If the solution exists for all $t \in I$ and $(u(t), v(t)) \in U$, then we can construct the functions $u : I \to \mathbb{R}$ and $v : I \to \mathbb{R}$ sending every $t \subset I$ to the values of the parameters u and v which solve system (3.14). In the case that $u = u(t)$ and $v = v(t)$ are regular functions, then

$$\tilde{\alpha} : I \to \mathbb{R}^3, \qquad \tilde{\alpha}(t) = X(u(t), v(t)), \quad t \in I$$

is associated to a regular parametrization $(I, \tilde{\alpha})$ of the curve $\alpha(I)$ which is contained in $X(U)$. Observe that the composition is a regular function. Moreover, for every $(x_0, y_0, z_0) \in \alpha(I)$ there exists $t_0 \in I$ such that $\alpha(t_0) = (x_0, y_0, z_0)$. Let $\{u_0, v_0\}$ be the solution of (3.14) with $t = t_0$ and $(u_0, v_0) \in U$. Then

$$X(u_0, v_0) = (x(u_0, v_0), y(u_0, v_0), z(u_0, v_0))$$

$$= (\alpha_1(t_0), \alpha_2(t_0, \alpha_3(t_0)) = (x_0, y_0, z_0).$$

This means that $(x_0, y_0, z_0) \in X(U)$.

The step of solving the system (3.14) for every $t \in I$ may be a hard (or even impossible) task, even locally in the parameters (u, v). In some cases, a direct inspection of the problem might be sufficient to solve the system. Other situations may be solvable with the help of stronger theories such as the Gröbner basis, used in symbolic computation during the last fifty years.

Example 3.4.1 We consider the cylinder parametrized by $(\mathbb{R}^2, X)$, where

$$X(u, v) = (\frac{1 - u^2}{1 + u^2}, \frac{2u}{1 + u^2}, v), \quad (u, v) \in \mathbb{R}^2.$$

Let $(\mathbb{R}, \alpha)$ be the space curve

$$\alpha(t) = \left(-\frac{2t^2 - 2t}{2t^2 - 2t + 1}, \frac{4t^2 - 2}{4t^2 - 4t + 2}, t + 1\right), \quad t \in \mathbb{R}.$$

The system (3.14) is determined by

$$\begin{cases} \dfrac{1 - u^2}{1 + u^2} = -\dfrac{2t^2 - 2t}{2t^2 - 2t + 1} \\ \dfrac{2u}{1+u^2} = \dfrac{4t^2-2}{4t^2-4t+2} \\ v = t + 1 \end{cases}$$

whose solutions in (u, v) are given by

$$u(t) = 2t - 1, \quad v(t) = t + 1.$$

Therefore, the curve $(\mathbb{R}, \alpha)$ is contained in the cylinder, and it holds that $X(u(t), v(t))$ is the parametrization of the curve. Figure 3.24 illustrates the example.

In the case that the surface is determined via a parametrization, and the curve is parametrized, a system of two equations in the two parameters describing the surface would appear, leading to a parametric solution if the curve is contained in the surface. Otherwise, if the surface is implicitly determined, the compatibility of an equation for all values of the parameter in the parametrization of the curve would determine that the curve is contained in the surface.

Finally, if both elements are determined by implicit equations, the compatibility condition is given by a system of three equations (two of them determining the curve) and three variables that the curve should satisfy. As mentioned above, this might be a hard problem.

When starting from a curve contained in a surface (U, X), we would wish the curve to be parametrized by $t \mapsto X(u(t), v(t))$, for some regular functions $u, v :$ $I \to \mathbb{R}$, for some open interval $I \subseteq \mathbb{R}$.

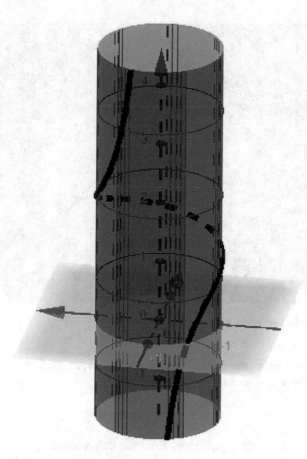

Fig. 3.24 Curve in surface

Example 3.4.2 We approximate each of the helicoidal ramps in the dome of Reichstag building (see Fig. 3.25) to fit the hemisphere

$$\{(x, y, z) \in \mathbb{R}^3 : x^2 + y^2 + z^2 = 1, z \geq 0\}.$$

In Sect. 2.5, we have modified the circular helix to satisfy certain necessities. Here, we construct a helix-like space curve lying in the hemisphere. The behavior of the curve is analogous to that of the helix, so we fix

$$x(t) = c_1(t) \cos(c_2(t)t), \quad y(t) = c_1(t) \sin(c_2(t)t), \quad z(t) = t, \qquad (3.15)$$

for some functions c_1, c_2 and all $t > 0$. We may assume that the turning rate regarding the parameter is constant in height so $c_2(t) = c_2$ for all $t > 0$. Moreover,

Fig. 3.25 Reichstag Dome, Berlin, by Foster+Partners

the third component of the parametrization of the curve, $z(t)$ should satisfy the equation of the surface, so it holds that

$$x(t)^2 + y(t)^2 + z(t)^2 = 1$$

for every $t > 0$. Regarding (3.15) and from the trigonometric properties, one has that

$$c_1 = \sqrt{1 - t^2},$$

which only makes sense for $0 \leq t \leq \beta \leq 1$. The value of β fixes the maximum height of the spiraling ramp in the dome. Varying the value of c_2 provides more turns of the curve before reaching the north pole. We write its parametrization (I, α_1), with

$$\alpha_1(t) = (\sqrt{1 - t^2}\cos(c_2 t), \sqrt{1 - t^2}\sin(c_2 t), t), \quad 0 < t < \beta.$$

A second ramp is rotated a half turn with respect to the previous one, i.e., it is parametrized by

$$\alpha_2(t) = (\sqrt{1 - t^2}\cos(c_2(t + \pi)), \sqrt{1 - t^2}\sin(c_2(t + \pi)), t), \quad 0 < t < \beta.$$

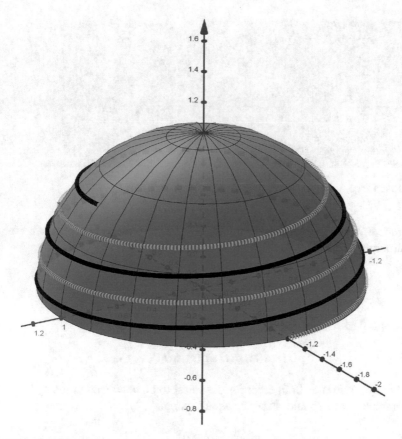

Fig. 3.26 Approximation of the ramps in the dome of Reichstag building

We refer to Sect. 2.5, where we have dealt with this kind of transformation. We fix $\beta = 0.68$ and $c_1 = (4\pi + \pi/2)/\beta$ to approximate the model. Figure 3.26 shows this approximated model.

Example 3.4.3 The ramp ascending the observation tower Camp Adventure in a forest near Copenhagen, Denmark, consists of a helicoidal ramp in a one-sheet hyperboloid. It was designed by Effekt. The structure of the tower can be found in detail in their webpage (effekt 2020a). Figure 3.27 shows the structure of such tower. For the sake of simplicity, we may assume that the hyperboloid is given by

$$\frac{x^2}{a^2} + \frac{y^2}{b^2} - \frac{z^2}{c^2} = 1,$$

Fig. 3.27 Camp Adventure Tower, in Haslev, Denmark, by Effekt

for some $a, b, c \in \mathbb{R}$ which can be obtained from the projections. More precisely, in the structure under study, we can approximate

$$a = b = 7.5m, \quad c = \frac{45}{2\sqrt{3}}m.$$

An spiral traversing the hyperboloid can be parametrized by

$$(\rho(t)\cos(t), \rho(t)\sin(t), h(t)), \quad t \in \mathbb{R},$$

for some $\rho(t), h(t) > 0$. In order for the spiral to be contained in the hyperboloid, its parametrization should satisfy its equation, i.e.,

$$\frac{\rho(t)^2}{a^2} - \frac{(h(t))^2}{c^2} = 1,$$

which means that

$$\rho(t) = a\cosh(s(t)), \quad h(t) = c\sinh(s(t)),$$

for some increasing function $s = s(t)$. Equivalently, the parametrization of the spiral is given by

$$(a\cosh(s)\cos(t(s)), b\cosh(s)\sin(t(s)), c\sinh(s)), \quad s \in \mathbb{R}.$$

In order to determine the function $s(t)$, we take into account that we aim to have equidistant levels in height. Therefore, $\sinh(s(t + 2\pi p)) - \sinh(s(t + 2\pi(p - 1)))$ should not depend on p. Approximating with a Taylor expansion at 0 the hyperbolic sinus we get that, in case we choose $s(t) = mt$, for some $m > 0$, this independence is attained. Moreover, the twelve turns in the 45 meter height of the tower determine a unique value of m.

Another curve contained in a sphere is **Seiffert's spiral** (see Erdös (2000)). It consists in a curve drawn by moving a point on the surface of a sphere, with fixed constant speed and angular velocity traversing from one point of the sphere to its opposite point in the sphere. Let $a \in \mathbb{R}$. It is straightforward to verify that the parametrization $(\mathbb{R}, \alpha)$ given by

$$\alpha(t) = \left(\frac{\cos(t)}{\sqrt{1 + (at)^2}}, \frac{\sin(t)}{\sqrt{1 + (at)^2}}, \frac{-at}{\sqrt{1 + (at)^2}} \right),$$

describes a curve contained in the unit sphere. This curve is known as a **spherical spiral**, or rhumb line. Many other curves embedded in a sphere, known as spherical curves, can be found in Mathcurve (2019a).

Another point of study concerning curves in surfaces is that of **geodesic curves**. These curves are associated to a surface and have interesting and practical properties. The geodesic joining two points in a surface can be defined as a curve contained in the surface, passing through the given points, and such that the length of the path between them is minimal. In the case where the surface is developable (see Sect. 4.7), the solution of the problem is quite intuitive, as stated in Pottmann et al. (2007), but the problem becomes more difficult in other regular surfaces.

It can be proved (see do Carmo (1976)) that a regular curve (I, α) is a geodesic of a surface if and only if $\alpha''(t)$ is collinear with the normal vector of the surface at the point $\alpha(t)$, for every point of the curve (see Costa et al. (1997)). Observe that this means that the velocity vector of the curve is constant, so the curve is traversed at constant speed. This concept can be described in terms of the elements obtained from the curvature of a surface.

Example 3.4.4 Let us consider the unit sphere. Every circumference of maximum radius is a geodesic curve contained in the unit sphere. Due to the symmetry of the sphere one may assume, without loss of generality, that the circumference is located at the floor plane. Therefore,

$$\alpha(t) = (\cos(t), \sin(t), 0), \quad t \in (0, 2\pi),$$

is a natural parametrization of such curve, which means

$$\alpha''(t) = (-\cos(t), -\sin(t), 0),$$

for all t. Observe that $-\alpha''(t)$ is a normal vector to the unit sphere at the same point.

Let us now give a few words about the study of intersection of surfaces. This problem has been studied from both numerical and symbolic points of view. We refer to Hoffmann (1989) and the references therein for the bases of some numerical and algorithmic techniques. Among them, we distinguish substitution maps. Generally speaking, given a surface S_1 in implicit form, say associated to $F(x, y, z) = 0$, and a surface S_2 in parametric form, say associated to the parametrization (U, X), with

$$X(u, v) = (x(u, v), y(u, v), z(u, v)), \quad (u, v) \in U,$$

the intersection curve could be obtained by searching for the points in the plane curve determined in implicit form by

$$F(x(u, v), y(u, v), z(u, v)) = 0.$$

Given any point (u_0, v_0) belonging to such curve, it holds that $X(u_0, v_0)$ is a point in the intersection of the two surfaces.

Other methods are based on substitution maps, projection and desingularization, least-squares approach, etc. We cite the paper (Narváez-Rodríguez et al. 2014), in which Roberto Narváez-Rodríguez, Andrés Martín-Pastor and María Aguilar-Alejandre study intersection of surfaces, quadrics, and more precisely, of cones.

In what follows, we will only make use of elementary techniques when intersecting surfaces in a very specific position.

Example 3.4.5 Let us consider a groin vault, created from the intersection of two barrel vaults. From the mathematical point of view, the intersection of such surfaces corresponds to the intersection of two cylinders crossing orthogonally. For simplicity, let us assume that both cylinders are

$$S_1 = \{(x, y, z) \in \mathbb{R}^3 : x^2 + z^2 - 1 = 0\}, \quad S_2 = \{(x, y, z) \in \mathbb{R}^3 : y^2 + z^2 - 1 = 0\}.$$

The intersection $S_1 \cap S_2$ defines two curves, each of which can be parametrized by $((0, 2\pi), \alpha_\pm)$, with

$$\alpha_\pm(t) = (\pm \cos(t), \cos(t), \sin(t)), \quad t \in (0, 2\pi).$$

Therefore, the intersection describes two ellipses. One of them is contained in the plane $x = y$ and the other in the plane $x = -y$. Their reduced equation is $\frac{(x^{\star\star})^2}{8} + (y^{\star\star})^2 = 1$. Figure 3.28 illustrates this intersection.

Example 3.4.6 Let us consider the sphere $S_1 = \{(x, y, z) \in \mathbb{R}^3 : x^2 + y^2 + z^2 - 4 = 0$ and the cylinder $S_2 = \{(x, y, z) \in \mathbb{R}^3 : (x - 1)^2 + y^2 - 1 = 0\}$. Regarding the cylinder, we have that in the coordinates XY, one can write $x - 1 = \cos(t)$, $y = \sin(t)$ for $t \in (0, 2\pi)$. Substituting this into the equation of the sphere yields

$$\sin^2(t) + z^2 + (\cos(t) + 1)^2 - 4 = 0,$$

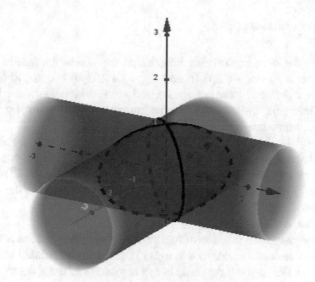

Fig. 3.28 Intersection of two cylinders, the basis of a groin vault

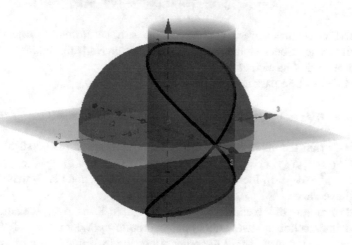

Fig. 3.29 Vibiani curve as the intersection of surfaces

or equivalently $z^2 = 2 - 2\cos(t)$. From the trigonometric relation $2\sin^2(t/2) = 1 - \cos(t)$ we derive that $z = 2\sin(t/2)$. The intersection of the surfaces can be parametrized by $(\mathbb{R}, \alpha)$, with

$$\alpha(t) = (1 + \cos(t), \sin(t), 2\sin(t/2)), \quad t \in \mathbb{R}.$$

We recall that this is the parametrization of Viviani's curve (see (2.1)). The geometric situation is illustrated in Fig. 3.29.

3.5   Suggested Exercises

3.1. Consider the locus of the point P such that the sum of the square of distances
to the points $P_1 = (-\sqrt{2}, 0, 0)$ and $P_2 = (\sqrt{2}, 0, 0)$ is equal to 6. Determine
the surface defined by the previous description. Which geometric form does it
have? Determine the locus when substituting the word "sum" by "difference"
in the previous statement.

3.2. Let $f : \mathbb{R}^3 \to \mathbb{R}$ be the function $f(x, y, z) = \exp(x - y + z)$ for all $(x, y, z) \in$
$\mathbb{R}^3$. Verify that $f \in C^\infty(\mathbb{R}^3)$ and compute its derivatives of any order.

3.3. Verify that any polynomial $p : \mathbb{R}^3 \to \mathbb{R}$ is such that $p \in C^\infty(\mathbb{R}^3)$.

3.4. Consider the surface of the cone in Example 3.1.7. Compute the area of such
cone up to height $h > 0$.

3.5. Compute the area of a sphere.

3.6. Given the graph of a function of two variables as in Proposition 3.1.9, compute
a normal vector associated to it at every point of the associated surface.

3.7. Let $p > 0$. We consider the (not always regular!) surface S determined by

$$S = \{(x, y, z) \in \mathbb{R}^3 : |x|^p + |y|^p + |z|^p = 1\}.$$

Parametrize the surface and sketch the surface from its projections. Such
surfaces have been studied in the literature. See Nadenik (2005) for the details
on several of its elements.

3.8. Let $(\mathbb{R}, \alpha)$ be a parametrization of the space curve

$$\alpha(t) = (\exp(t) - \exp(-t), \exp(t) + \exp(-t), \exp(t)), \quad t \in \mathbb{R}.$$

Verify that the curve is contained in the intersection of the surfaces defined
by $S_1 = \{(x, y, z) \in \mathbb{R}^3 : xz - z^2 + 1 = 0\}$ and $S_2 = \{(x, y, z) \in \mathbb{R}^3 :$
$yz - z^2 - 1 = 0\}$. Identify the surfaces S_1 and S_2, and verify that the curve is
a plane curve.

3.9. The structure of the church Kópavogskirkja, in Kópavogur, Iceland, describes
the intersection of two orthogonal parabolic cylinders. One of them corresponds to the turn of $\pi/2$ the other. Describe the intersection of the surfaces.
An old photograph of the church is in Fig. 3.30.

Fig. 3.30 Church of Kópavogur, Iceland

Chapter 4
Special Families of Surfaces

In this last chapter we focus on some families of surfaces of great importance in practice. These are ruled surfaces, surfaces of revolution and quadrics. These families have nonempty intersections. In addition to that, we will also go into some detail on certain subfamilies which are of great importance in applications in architecture. As a matter of fact, more enriching examples of the theory can be found in works of architecture.

4.1 Ruled Surfaces

The study of ruled surfaces is a significant milestone for the applications of the theory of surfaces in architecture. Structural elements such as beams give rise to the appearance of this kind of surfaces in architectural models, providing stability to the structure. Among them, one can find a great variety of examples satisfying a wide range of desired properties, not only stability but aesthetic, physical, etc.

In this section, we only give a brief overview on the topic, before applying these concepts to specific architectural elements in Sects. 4.2 and 4.3.

Definition 4.1.1 A **ruled surface** is a surface S determined by the movement of a line in space. Such surfaces can be parametrized by (U, X), with $U = I \times \mathbb{R}$ for some nonempty open interval $I \subseteq \mathbb{R}$, with

$$X(u, v) = \alpha(u) + v\omega(u), \quad (u, v) \in I \times \mathbb{R},$$

where (I, α) corresponds to the parametrization of a space curve, and $\omega(u)$ is a non-zero three-dimensional vector, for each $u \in I$.

Observe from the previous definition that for each $u \in I$, the line at $\alpha(u)$ and vector $\omega(u)$ is totally contained in the ruled surface. Each of these lines are known

© The Author(s), under exclusive license to Springer Nature Switzerland AG 2021
A. Lastra, *Parametric Geometry of Curves and Surfaces*, Mathematics and the Built
Environment 5, https://doi.org/10.1007/978-3-030-81317-8_4

Fig. 4.1 A ruled surface and some of its generators

as a **generator** of the ruled surface. The space curve $\alpha(I)$ is also totally contained in S, and it is known as the **directrix** of the ruled surface.

Observe that not every ruled surface is a regular surface, as can be deduced from the cone in Example 3.1.7. A less conventional ruled surface is displayed in Fig. 4.1.

Among all the ruled surfaces there are many different subcategories which are worth studying separately. We only go into detail on some of them. It is worth remarking that all of them belong to a wider class of surfaces known as **developable surfaces**. The main property of the elements in this class is that they can be locally flattened onto a plane with no stretching nor tearing. From the previous condition, we have that any developable surface must be a ruled surface (Pottmann and Wallner 2001), but the two classes do not coincide. Not aiming to remain exhaustive in this definition, we only state a mathematical characterization of a developable surface: ruled surfaces which has the same tangent plane at all points along a generator.

Conical surfaces, cylindrical surfaces and tangent developable surfaces are examples of developable surfaces. However, there are more exotic examples such as the **oloid** and the **sphericon**.

A parametrization of the oloid is

$$\left(\sin(u)v, \left(-\frac{1}{2} - \cos(u)\right)v + \left(\frac{1}{2} - \frac{\cos(u)}{1 + \cos(u)}\right)(1 - v), \pm\frac{\sqrt{1 + 2\cos(u)}}{1 + \cos(u)}(1 - v)\right),$$

for $u \in (0, 2\pi)$, $v \in (0, 1)$. An illustration of the oloid can be found in Fig. 4.2, where the rules for its construction become clear. An oloid is obtained by joining two orthogonal circles by segments. The QR Code of Fig. 4.2 links to different views of this surface.

Regarding the spheroid, this surface consists in the ruled surface joining points in two semicircles. It can be parametrized by the following regular parametrizations:

$$(\cos(u)v, \sin(u)v, \pm(1 - v)), \quad u \in (-\pi, 0), v \in (0, 1),$$

$$(\pm(1 - v), \cos(u)v, \sin(u)v), \quad u \in (-\pi/2, \pi/2), v \in (0, 1).$$

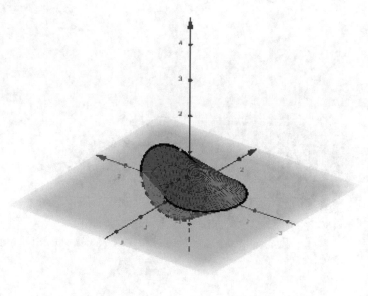

Fig. 4.2   Oloid. QR Code 17

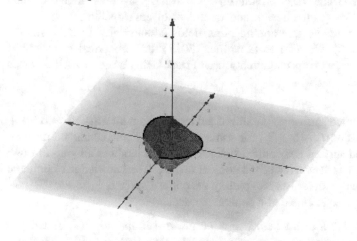

Fig. 4.3   Sphericon. QR Code 18

An illustration of the sphericon can be found in Fig. 4.3, where the rules for its construction become clear. The QR code in Fig. 4.3 links to different views of this surface.

There is an increasing interest in developable surface due to the effects of the light on them and many other features. In Glaeser and Gruber (2007), Georg

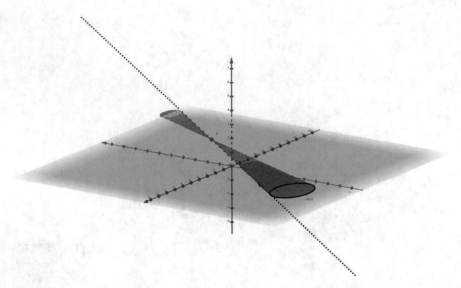

Fig. 4.4 Conical surface with an ellipse at height $z = -2$ as directrix, and vertex $(1, 1, 2)$

Glaeser and Franz Gruber discuss buildings which have taken advantage of this. Some recent methods for the design and treatment of developable structures and their application in architecture among other fields is considered in Krivoshapko and Shambina (2012). We refer to Lawrence (2011) for a historical overview of developable surfaces, with special attention on Frank Gehry's use of such surfaces in some of his works.

Conical Surface

Assume that in a ruled surface S, parametrized by (U, X) with $X(u, v) = \alpha(u) + v\omega(u)$, it holds that $\omega(u) = \overrightarrow{\alpha(u)}P$, for some fixed $P \in \mathbb{R}^3$ not contained in $\alpha(I)$. The resulting ruled surface is a conical surface whose generatices all pass through the point P, known as the **vertex** of the conical surface. The knowledge of both the vertex and the directrix describes completely the conical surface.

Figure 4.4 shows an example of a conical surface.

Proposition 4.1.2 *Let a conical surface S be given with directrix (I, α) being a regular space curve, and vertex P, satisfying that the set $\{\alpha'(u), \overrightarrow{\alpha(u)}P\}$ is a linearly independent set of vectors, for every $u \in I$. Then the only singular point of S is its vertex.*

More precisely, the singular points of S is the set

$$\{P\} \cup \bigcup_{u \in I, v \in \mathbb{R}} \{\alpha(u) + v\overrightarrow{\alpha(u)}P : \alpha'(u) \parallel \overrightarrow{\alpha(u)}P\}.$$

Proof Let us consider the parametrization of a conical surface (U, X) where $U = I \times \mathbb{R}$, and assume that $(I, \alpha = (\alpha_1, \alpha_2, \alpha_3))$ is the directrix associated to the conical surface and $P = (x_0, y_0, z_0)$ is its vertex. Topological arguments are sufficient in order to guarantee that any neighborhood of the vertex in the conical surface is disconnected when removing the vertex, so it can not be homeomorphic to a plane. For each of the other points in the conic surface there exists a neighborhood of the point in the cone which is homeomorphic to a plane.

Regarding the condition on the rank in Definition 3.1.2 we observe that $X(u, v) = \alpha(u) + v\alpha(\vec{u})P$ determines the matrix

$$\begin{pmatrix} \alpha_1'(u)(1-v) & \alpha_2'(u)(1-v) & \alpha_3'(u)(1-v) \\ \alpha_1(u) - x_0 & \alpha_2(u) - y_0 & \alpha_3(u) - z_0 \end{pmatrix}^T.$$

The three minors of order 2 of the previous matrix are given by

$$(\alpha_2(u) - y_0)\alpha_1'(u)(1-v) - \alpha_2'(u)(1-v)(\alpha_1(u) - x_0),$$

$$(\alpha_3(u) - z_0)\alpha_1'(u)(1-v) - \alpha_3'(u)(1-v)(\alpha_1(u) - x_0),$$

$$(\alpha_3(u) - z_0)\alpha_2'(u)(1-v) - \alpha_3'(u)(1-v)(\alpha_2(u) - y_0). \tag{4.1}$$

The three of them are 0 in case $v = 1$, i.e., at the vertex. The curve (I, α) is regular. Therefore, for all $u \in I$ there exists $j = 1, 2, 3$ such that $\alpha_j'(u) \neq 0$. Let $u \in I$ and assume that $\alpha_j'(u) \neq 0$ for $j = 1, 2, 3$. All the expressions in (4.1) are null whenever

$$\frac{\alpha_1(u) - x_0}{\alpha_1'(u)} = \frac{\alpha_2(u) - y_0}{\alpha_2'(u)} = \frac{\alpha_3(u) - z_0}{\alpha_3'(u)},$$

The previous condition is attained if and only if there exists $\lambda \in \mathbb{R}^*$ such that

$$\alpha'(u) = (\alpha_1'(u), \alpha_2'(u), \alpha_3'(u)) = \lambda(\alpha_1(u) - x_0, \alpha_2(u) - y_0, \alpha_3(u) - z_0)$$

$$= \lambda\alpha(\vec{u})P. \tag{4.2}$$

In case that one of the components of the velocity vector at $u \in I$ vanishes, for instance $\alpha_1'(u) = 0$, then we have from the two first conditions in (4.1) one of the following:

- $\alpha_1(u) = x_0$, in which case (4.2) holds for the first component. If the other components do not vanish, then (4.2) holds. Otherwise, one has two possibilities:
  - $\alpha_2'(u) = 0$, which means that $\alpha_2(u) = y_0$, from the third element in (4.1).
  - or $\alpha_3'(u) = 0$, which yields $\alpha_3(u) = z_0$.
- or $\alpha_2'(u) = \alpha_3'(u) = 0$, which is not a feasible situation.

This completes all the possible cases. In conclusion, the singular points of S are its vertex and the points in the line joining P and $\alpha(u)$ whenever $\alpha'(u) \| \alpha(\vec{u})P$.

Example 4.1.3 Let $(I = (0, 2\pi), \alpha)$ be the parametrization of the space curve

$$\alpha(u) = (\cos(u), \cos^2(u), \sin(u)), \quad u \in (0, 2\pi),$$

and let $P = (4, 0, 1)$. The conical surface with directrix (I, α), and vertex P is parametrized by

$$X(u, v) = (\cos(u)(1 - v) + 4v, \cos^2(u)(1 - v), \sin(u)(1 - v) + v)$$

$$= (x(u, v), y(u, v), z(u, v)),$$

for $(u, v) \in (0, 2\pi) \times \mathbb{R}$. Observe that for every $(u_0, v_0) \in (0, 2\pi) \times \mathbb{R}$ we have

$$\text{rank} \left(\begin{matrix} \frac{\partial x}{\partial u}(u_0, v_0) & \frac{\partial y}{\partial u}(u_0, v_0) & \frac{\partial z}{\partial u}(u_0, v_0) \\ \frac{\partial x}{\partial v}(u_0, v_0) & \frac{\partial y}{\partial v}(u_0, v_0) & \frac{\partial z}{\partial v}(u_0, v_0) \end{matrix} \right)^T$$

$$= \text{rank} \left(\begin{matrix} -(1 - v_0)\sin(u_0) & -2\sin(u_0)\cos(u_0)(1 - v_0) & \cos(u_0)(1 - v_0) \\ 4 - \cos(u_0) & -\cos^2(u_0) & 1 - \sin(u_0) \end{matrix} \right)^T = 1$$

whenever $v_0 = 1$, i.e., at the vertex of the conical surface, and also if

$$\cos^2(u_0)\sin(u_0) + 8\cos(u_0)\sin(u_0) = 2\cos^2(u_0)\sin(u_0)$$

$$\sin^2(u_0) - \sin(u_0) - 4\cos(u_0) + \cos^2(u_0) = 0$$

$$2\sin^2(u_0)\cos(u_0) + \cos^3(u_0) = 2\sin(u_0)\cos(u_0)$$

hold simultaneously. Actually, the three equations can only hold for $u_0 = \pi/2$, and any $v_0 \in \mathbb{R}$, so the whole line passing through $(0, 0, 1)$ and the vertex is a line consisting of singular points. Figure 4.5 displays this situation.

Cylindrical Surface
Cylindrical surfaces satisfy the condition that the vector $\omega(u)$ does not depend on u, so any of these surfaces can be parametrized by

$$X(u, v) = \alpha(u) + v\omega, \quad (u, v) \in I \times \mathbb{R},$$

where (I, α) is the directrix of the ruled surface, and $\omega \in \mathbb{R}^3 \setminus \{0\}$.

For example, the cylindrical surface of directrix given by one branch of a hyperbola at height $z = 0$, $\omega = (0, 3, 3)$ is illustrated in Fig. 4.6.

A result analogous to Proposition 4.1.2 can be stated for cylindrical surfaces.

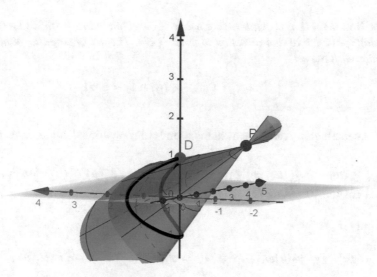

Fig. 4.5 Conical surface in Example 4.1.3

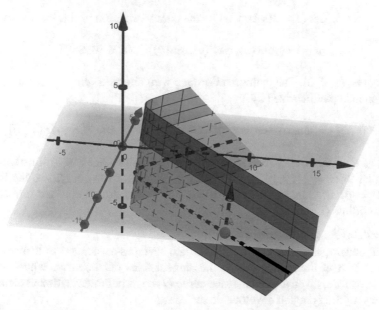

Fig. 4.6 Cylindrical surface of directrix given by one branch of a hyperbola at height $z = 0$, $\omega = (0, 3, 3)$

Proposition 4.1.4 *Let a cylindrical surface S be given with directrix (I, α) being a regular space curve and direction $\omega \in \mathbb{R}^3 \setminus \{0\}$. The set of singular points of S coincides with the set*

$$\bigcup_{v \in \mathbb{R}, u \in I} \{\alpha(u) + v\omega : \alpha'(u) \parallel \omega, t \in \mathbb{R}\}.$$

Proof An analogous reasoning as in the proof of Proposition 4.1.4 leads us to

$$\text{rank} \begin{pmatrix} \frac{\partial x}{\partial u}(u_0, v_0) & \frac{\partial y}{\partial u}(u_0, v_0) & \frac{\partial z}{\partial u}(u_0, v_0) \\ \frac{\partial x}{\partial v}(u_0, v_0) & \frac{\partial y}{\partial v}(u_0, v_0) & \frac{\partial z}{\partial v}(u_0, v_0) \end{pmatrix}^T = \text{rank} \begin{pmatrix} \alpha_1'(u) & \alpha_2'(u) & \alpha_3'(u) \\ \omega_1 & \omega_2 & \omega_3 \end{pmatrix}^T \neq 2$$

if and only if

$$\alpha_1'(u)\omega_2 = \alpha_2'(u)\omega_1, \quad \alpha_1'(u)\omega_3 = \alpha_3'(u)\omega_1, \quad \alpha_2'(u)\omega_3 = \alpha_3'(u)\omega_2$$

where $\alpha'(u) = (\alpha_1'(u), \alpha_2'(u), \alpha_3'(u))$, and $\omega = (\omega_1, \omega_2, \omega_3)$. The conclusion follows directly from here.

Example 4.1.5 Let $(I = (0, 2\pi), \alpha)$ be the parametrization of the space curve

$$\alpha(u) = (\cos(u), \cos^2(u), \sin(u)), \quad u \in (0, 2\pi),$$

and let $\omega = (1, 0, 0)$. The cylindrical surface with directrix (I, α), associated to the direction ω is parametrized by

$$X(u, v) = (\cos(u) + v, \cos^2(u), \sin(u)) = (x(u, v), y(u, v), z(u, v)),$$

for $(u, v) \in (0, 2\pi) \times \mathbb{R}$. Observe that $\alpha'(u) = (-\sin(u), -2\cos(u)\sin(u), \cos(u))$ is parallel to $\omega = (1, 0, 0)$ if and only if $u = \pi/2$ or $u = 3\pi/2$, so the lines at $\alpha(\pi/2) = (0, 0, 1)$ and direction ω and $\alpha(3\pi/2)$ and direction ω are formed with singular points (Fig. 4.7).

Tangent Developable Surface

Tangent developable surfaces are those ruled surfaces associated to a space curve which consist of the lines given by the tangent lines of the curve, whenever they exist. Therefore, given a regular space curve (I, α), a parametrization associated to its corresponding tangent developable surface is

$$X(u, v) = \alpha(u) + v\alpha'(u), \quad (u, v) \in I \times \mathbb{R}.$$

Example 4.1.6 An example of such surface is the following. Let us consider the circular helix parametrized by

$$\alpha(u) = (\cos(u), \sin(u), u), \quad u \in \mathbb{R}.$$

Fig. 4.7 The cylindrical surface in Example 4.1.5

Fig. 4.8 The tangent developable surface in Example 4.1.6

Its associated tangent developable surface is described by

$$X(u, v) = \alpha(u) + v\alpha'(u)$$
$$= (\cos(u) - v\sin(u), \sin(u) + v\cos(u), u + v), \quad (u, v) \in \mathbb{R}^2.$$

Figure 4.8 shows the shape of that surface.

Proposition 4.1.7 *Let a tangent developable surface S be given, with directrix (I, α) a regular space curve. Then the set of singular points of S is determined by the set*

$$\alpha(I) \cup \{\alpha(u) + v\alpha'(u) : \alpha(u) \text{ is an inflection point of } (I, \alpha)\}.$$

Proof *Any point of the tangent developable surface is defined by $\alpha(u) + v\alpha'(u)$, for some $u \in I$, $v \in \mathbb{R}$. One has that*

$$\text{rank} \begin{pmatrix} \alpha'_1(u) + v\alpha''_1(u) & \alpha'_2(u) + v\alpha''_2(u) & \alpha'_3(u) + v\alpha''_3(u) \\ \alpha'_1(u) & \alpha'_2(u) & \alpha'_3(u) \end{pmatrix}$$

$$= \text{rank} \begin{pmatrix} v\alpha''_1(u) & v\alpha''_2(u) & v\alpha''_3(u) \\ \alpha'_1(u) & \alpha'_2(u) & \alpha'_3(u) \end{pmatrix}.$$

If $v = 0$, then the above rank equals 1. This holds for all the points of the surface lying in the directrix. If $v \neq 0$, then the rank is less than 2 in case $\alpha'(u)$ and $\alpha''(u)$ are proportional, i.e., $T_\alpha(u). \|.T'_\alpha(u)$, or in terms of Lemma 2.2.13, $\alpha(u) + v\alpha'(u)$ is such that $\alpha(u)$ is an inflection point of (I, α).

4.2 Some Subfamilies of Ruled Surfaces

In this section, we go into some detail about certain surfaces of interest for applications, and whose structures often appear in CAD programs.

Catalan Surfaces and Spiral Staircases

A **Catalan surface** is a ruled surface (see Sect. 4.1) whose generators are such that their directions form a vector space of dimension 2, i.e., all of them are contained in a plane. More precisely, a Catalan surface can be parametrized by (U, X) for $U = I \times \mathbb{R}$, with I being an open interval, and such that

$$X(u, v) = \alpha(u) + v\omega(u), \quad (u, v) \in I \times \mathbb{R},$$

with

$$L = \{\omega(u) : u \in I\} \subseteq \mathbb{R}^3$$

being a subspace of dimension 2. The previous condition can be rewritten in the form

$$\omega(u) = \lambda_1(u)\omega_1 + \lambda_2(u)\omega_2$$

for some fixed linearly independent vectors ω_1, $\omega_2 \in \mathbb{R}^3$. The elements $\lambda_1(u)$, $\lambda_2(u)$ are real numbers which determine the coordinates of the vector $\omega(u)$ in the basis of the subspace of dimension 2 generated by the set of vectors $\{\omega_1, \omega_2\}$.

In Dzwierzynska and Prokopska (2018), Jolanta Dzwierzynska and Aleksandra Prokopska develop a novel approach in parametric design of roof shells formed by the repetition of Catalan surfaces. In Tofil (2007), a didactic approach to roof design through Catalan surfaces is shown.

One of such surfaces is the **helicoid**. The canonical helicoid can be parametrized by (U, X), where $U = \mathbb{R} \times \mathbb{R}$. In the parametrization $X(u, v) = \alpha(u) + v\omega(u)$, the directrix $(\mathbb{R}, \alpha)$ is the vertical line of direction $(0, 0, 1)$, and for every $u \in \mathbb{R}$, the generatrices are given by the vector $(\cos(u), \sin(u), 0)$. Observe that the vector space generated by these vectors is of dimension 2. All of them are contained in the vector plane $z = 0$. Therefore, the parametrization of a helicoid reads as follows:

$$X(u, v) = (v\cos(u), v\sin(u), u), \quad (u, v) \in \mathbb{R}^2.$$

For practical purposes, it is usual to consider not the whole generatrices, but an interval $u \in (0, a)$, for some $a > 0$. As in Sect. 2.5, one can deform this surface by adequate modifications in order to satisfy certain necessities. For example, one can consider the helicoid parametrized by

$$X(u, v) = (v\cos(u), v\sin(u), cu), \quad (u, v) \in \mathbb{R} \times (0, a), \tag{4.3}$$

for some fixed $c > 0$ (see Fig. 4.9).

Fig. 4.9 Helicoid. $c = 0.2$

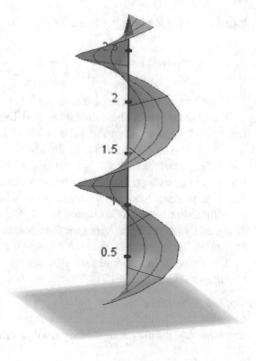

Generalized helicoids modify the previous construction in such a way that the curve defined in the parameter v is some curve other than a line. For example, given the space curve parametrized by $(\alpha_1(v), \alpha_2(v), \alpha_3(v))$, for $v \in I$, the generalized helicoid is parametrized by

$$X(u, v) = (x(u, v), y(u, v), z(u, v))$$

where

$$\begin{pmatrix} x(u, v) \\ y(u, v) \\ z(u, v) \end{pmatrix} = \begin{pmatrix} \cos(u) & -\sin(u) & 0 \\ \sin(u) & \cos(u) & 0 \\ 0 & 0 & 1 \end{pmatrix} \begin{pmatrix} \alpha_1(v) \\ \alpha_2(v) \\ \alpha_3(v) \end{pmatrix} + \begin{pmatrix} 0 \\ 0 \\ u \end{pmatrix} \quad (4.4)$$

for $v \in I$, $u \in \mathbb{R}$. Observe from Eq. (4.4) that the particular case of the helicoid in Eq. (4.3) is such that the curve (I, α) is the segment $\alpha(v) = (v, 0, 0)$, for $I = (0, a)$. It is also worth mentioning that Eq. (4.4) consists on the rotation of the space curve $\alpha(I)$ (see Sect. 4.4) while it is being translated upwards.

An example of these kinds of curves are displayed in Fig. 4.11, for the lemniscate (see Sect. 1.2):

$$\beta(v) = \left(\frac{2\cos(v)}{\sin^2(v) + 1}, \frac{\sin(v)\cos(v)}{\sin^2(v) + 1} \right), \quad v \in \mathbb{R},$$

located in the XZ plane, i.e., parametrized by

$$\beta(v) = \left(\frac{2\cos(v)}{\sin^2(v) + 1}, 0, \frac{\sin(v)\cos(v)}{\sin^2(v) + 1} \right), \quad v \in \mathbb{R}.$$

Therefore, the parametrization of such surface is given by Eq. (4.4).

The movement of the lemniscate and the resulting surface can be observed following the QR Code in Fig. 4.10.

It is natural to find helicoids in the form of staircases (see Fig. 1.11), not only in simple form, but also in pairs, one shifted with respect to the other, which allows the user to go up or down the stair without meeting anyone going in the other direction. This is the case of a double staircase in the Vatican Museum (see Fig. 4.11). We also mention Other very famous examples of double staircases are the double staircase designed by Leonardo da Vinci for the Chateau de Chambord in France, and the one in the Reichstag Dome in Berlin by Foster + Partners discussed in Chap. 2.

The shape of a helicoid has also appeared in architectural elements, such as skyscrapers (Capanna 2012).

Conoid Structures
Conoids form a subclass of Catalan surfaces. In addition, these surfaces satisfy the condition that all the generatrices pass through a given line.

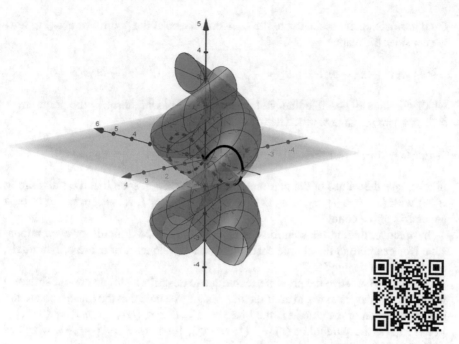

Fig. 4.10 Example of a generalized helicoid based on a lemniscate. QR Code 19

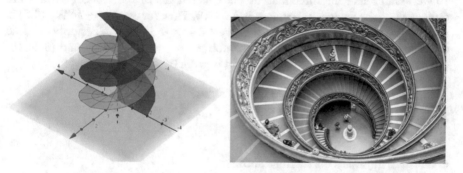

Fig. 4.11 Double lemniscate (left), and double staircase in Vatican Museum, Rome (right)

We proceed to describe how to come up with a parametrization of such surfaces in a general procedure, as follows. Assume that the directrix is determined by (I, α), with $\alpha = (\alpha_1, \alpha_2, \alpha_3)$. We write

$$(x, y, z) = (x_0, y_0, z_0) + (m_1, m_2, m_3)s, \quad s \in \mathbb{R},$$

for the line where all generatrices pass. Here, $(x_0, y_0, z_0) \in \mathbb{R}^3$ is a point of the line, and $(m_1, m_2, m_3) \neq (0, 0, 0)$ is the direction vector of the line. Let (n_1, n_2, n_3) be a fixed normal vector associated to the Catalan surface at the point. For every $t_0 \in I$, we consider the plane

$$n_1(x - x_0 - m_1t_0) + n_2(y - y_0 - m_2t_0) + n_3(z - z_0 - m_3t_0) = 0,$$

which contains all possible lines in the surface and passing through the point $(x_0 + m_1t_0, y_0 + m_2t_0, z_0 + m_3t_0)$. The solutions of the equation

$$n_1(\alpha_1(s) - x_0 - m_1t_0) + n_2(\alpha_2(s) - y_0 - m_2t_0) + n_3(\alpha_3(s) - z_0 - m_3t_0) = 0, \quad (4.5)$$

in $s \in I$ are the values of the parameter, say $\{s_1, \ldots, s_n\}$, such that the line joining $\alpha(s_j)$ with $(x_0 + m_1t_0, y_0 + m_2t_0, z_0 + m_3t_0)$ for $j = 1, \ldots, n$, turns out to be a generatrix of the conoid.

In practice, finding the solutions of Eq. (4.5) might be difficult, or even impossible (for example in the simple case where the components of α are polynomials, one of them of degree 5).

We put into practice the previous technique to describe a simple conoid structure which is of widespread use in architecture design. Assume that the line associated to the construction of the conoid is the line $x = z = 0$, i.e., $(m_1, m_2, m_3) = (0, 1, 0)$. Let (I, α) be the parabola $\alpha(s) = (1, s, -s^2)$, for $s \in I = (-1, 1)$, which is contained in the plane $x = 1$.

We assume that the directions of the generatrices of the conical surface are orthogonal to the vector $(n_1, n_2, n_3) = (0, 1, 0)$. Then for every $t = t_0$, Eq. (4.5) is reduced to $s - t_0 = 0$. Therefore, for every $-1 < s < 1$, there exists only one line contained in the conoid, which joins the points $\alpha(t_0) = (1, t_0, -t_0^2)$ and $(0, t_0, 0)$. Therefore, a parametrization of the conoid is given by

$$\begin{cases} x(t, s) = & 1 + s \\ y(t, s) = & t \\ z(t, s) = & t^2 + st^2 \end{cases}, \quad t \in (-1, 1), s \in \mathbb{R}.$$

Figure 4.12 illustrates the example above.

The previous example appears frequently in architectural realizations (see the suggested exercises at the end of this chapter).

Also, other curves such as trigonometric ones determine conoid structures in the Schools of Sagrada Familia by Gaudí (see Fig. 4.13). See Dolezal (2011) for more details and examples and a model of the previous surface.

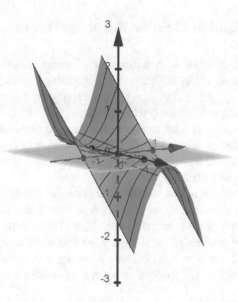

Fig. 4.12 Example of a parabolic conoid

Fig. 4.13 Sagrada Familia, Schools in Barcelona by Antoni Gaudí

4.3   Parametrization of Some Ruled Surfaces

In this section, we describe with some examples the technique of constructing certain ruled surfaces under additional interpolating conditions, such as their occurrence on two spatial curves. An in-depth study on these kind of surfaces is found in Krivoshapko and Ivanov (2015, Section 1.1.1).

Given two spatial curves with an empty intersection, parametrized by (I, α) and (J, β), we aim to design a ruled surface in which each curve is a directrix. This means that the surface emerges from a ruling. Each line is determined by two points, one in each of the space curves given.

A first question which arises is how to choose the points in order to form the surface. One can assume that the choice is associated to the way one traverses the curves, regarding their given parametrization. Therefore, a reparametrization of one of the curves can be performed via a scaling, so that both curves are defined in the same interval, say I. Let $I = (a_1, b_1)$ and $J = (a_2, b_2)$, with $a_j < b_j$ being real numbers or $\pm\infty$, for $j = 1, 2$. More precisely, we consider the map

$$f(t) = b_2 + \frac{b_1 - b_2}{a_1 - a_2}(t - a_2), \tag{4.6}$$

if $a_1, a_2, b_1, b_2 \in \mathbb{R}$. The case of any of them being $\pm\infty$ is not going to be considered in applications. However, it is worth remarking that in that situation, we may deal with families of transformations, such as

$$f_\lambda(t) = \lambda\frac{a_1 - b_1}{a_1 - t} + b_2 - \lambda,$$

for any fixed $\lambda > 0$, in the situation that $a_2 = -\infty$ and the other extreme points of the intervals are real numbers.

Observe from Eq. (4.6) that $f(I) = J$, and $(I, \beta \circ f)$ is a reparametrization of the space curve (J, β) with I being the novel interval of definition of the parameter. From now on, we will assume that both intervals coincide.

Among the realizations that one may consider in architectue, we analize Louis Kahn's Kimbell Art Museum in Ft. Worth, Texas (see Sect. 1.2 and Fig. 4.15), and also the church of San Juan de Ávila, in Alcalá de Henares, by Eladio Dieste (see Fig. 4.14). The structure of the walls studied in Rossi and Palmieri (2020) is also of great interest, as pointed out by Kim Williams. An in-depth explanation about the works of Eladio Dieste can be found in Anderson (2004), where other structures which might be interesting to describe by parametrization are mentioned, such as Robert Maillart's Zementhalle, or Cement Hall (Anderson 2004, p. 111).

For the sake of clarity, we will focus on the geometric technique, rather than the accuracy of measurement.

The first example to consider is the case of a cylindrical structure, i.e., when the surface is a cylindrical surface (see Sect. 4.1). A translation of a curve is performed in order to transform the curve $\alpha(I)$ into $\beta(I)$. Let us fix a vault shell

Fig. 4.14 The church of San Juan de Ávila in Alcalá de Henares, Spain, by Eladio Dieste

of $h = 2$ meters high, $w = 6$ meters wide and $\ell = 30.6$ meters long. The directrix determining the shell is a cycloid. We state the origin of coordinates in order that the cycloids are parametrized by

$$\alpha(u) = (h(u - \sin(u)), 0, h(1 - \cos(u))), \quad 0 < u < 2\pi,$$

$$\beta(u) = (h(u - \sin(u)), \ell, h(1 - \cos(u))), \quad 0 < u < 2\pi.$$

Therefore, the infinite cylinder is parametrized by

$$X_1(u, v) = \alpha(u) + v(0, 1, 0),$$

for $u \in (0, 2\pi) = I$, and $v \in \mathbb{R}$. In order to draw the part of the cylinder located between $\alpha(I)$ and $\beta(I)$ we put

$$X(u, v) = \frac{v}{\ell}\beta(u) + \frac{\ell - v}{\ell}\alpha(u), \quad 0 < u < \pi, 0 < v < \ell.$$

Observe that for all fixed u, the curve $X(u, v)$ determines the segment of endpoints $\alpha(u)$ and $\beta(u)$. For the value $u = 0$ we draw (I, α), and for $u = \ell$ the cycloid (I, β) appears. This does not enter into conflict with the given parametrization when considering the surface as cylindrical (see Sect. 4.1) due to

$$X(u, v) = \frac{v}{\ell}\beta(u) + \frac{\ell - v}{\ell}\alpha(u) = \alpha(u) + \frac{v}{\ell}(\beta(u) - \alpha(u)) = X_1(u, v),$$

$$\begin{cases} x(u, v) = \frac{v}{\ell}h(u - \sin(u)) + \frac{\ell-v}{\ell}h(u - \sin(u)) \\ y(u, v) = v - 1 \\ z(u, v) = \frac{v}{\ell}h(1 - \cos(u)) + \frac{\ell-v}{\ell}h(1 - \cos(u)) \quad 0 < u < 2\pi, 0 < v < \ell, \end{cases}$$

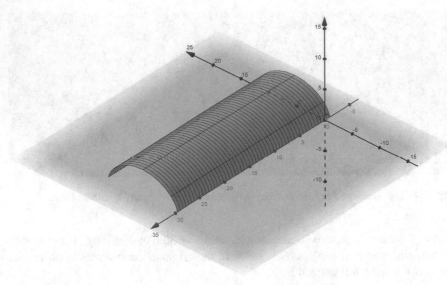

Fig. 4.15 The vault of the Kimbell Art Museum, by Louis Kahn

for all $0 < u < 2\pi$ and $v \in \mathbb{R}$. Figure 4.15 illustrates the cylinder constructed by the previous parametrization.

A second step is to consider different space curves determining the surface. As a first example, one can study the church of Cristo Obrero in Atlántida, Uruguay, also by Eladio Dieste (see Fig. 4.16).

Built in brickwork, its lateral walls determine a ruled surface which can be approximated as follows. At height $h = 6.75$, we fix the sinusoidal curves

$$\alpha_1(u) = (u, r\cos(\omega_1 u), h), \quad 0 < u < \ell_1,$$

and

$$\alpha_2(u) = (u, r\cos(\omega_1 u - \omega_2) - w, h), \quad 0 < u < \ell_1,$$

for $\ell = 31.61$, $w = 15.4$, $r = 1.4$ and $\omega_1 = 1.09$, $\omega_2 = \pi$. At the floor plane, we fix the lines

$$\beta_1(u) = (u, 0, 0), \quad 0 < u < \ell_1,$$

and

$$\beta_2(u) = (u, 0, -w), \quad 0 < u < \ell_1,$$

for the sake of simplicity.

Fig. 4.16 Detail of the Church of Cristo Obrero in Atlántida, Uruguay, by Eladio Dieste

It is worth remarking this configuration is closer to an approximation of the church of San Juan de Ávila, by the same architect. Indeed, the curve on the floor plane in what concerns us should stay at some positive height.

We construct the walls as the part of the ruled surface between the curves α_i and β_i, with segments with endpoints lying on the previous curves (Fig. 4.17).

The parametrization of each of the walls is given by

$$X_1(u, v) = \alpha_1(u)\frac{h - v}{h} + \beta_1(u)\frac{v}{h}, \quad 0 < u < \ell_1, 0 < v < h,$$

and

$$X_2(u, v) = \alpha_2(u)\frac{h - v}{h} + \beta_2(u)\frac{v}{h}, \quad 0 < u < \ell_1, 0 < v < h,$$

respectively.

Notice that an analogous procedure can be followed with Dieste's church of San Juan de Ávila. Another example is Maillart's Zementhalle, built in 1939 and demolished the next year. In this reinforced concrete structure one could distinguish a ruled surface between two arcs at different heights.

A step further in the parametrization of surfaces can be taken in this direction. So far, we have constructed a ruled surface from the endpoints determining the segments contained in the surface. One can also build surfaces by means of other

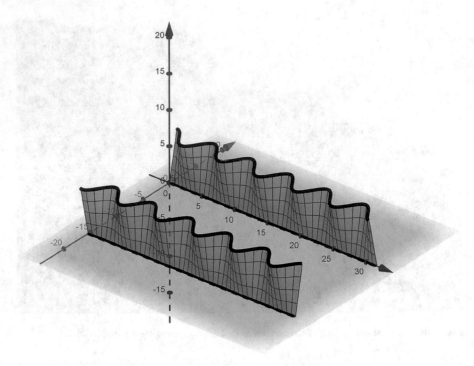

Fig. 4.17 Walls of the church of Cristo Obrero

more general curves passing through higher number of points. In the case that such points are in general position, the space curve fitting the interpolating points should be chosen of some specific nature. For example, the use of a masonry vault in the church of Cristo Obrero, with double curvature, is known to follow a Gaussian vault geometry. More precisely, this vault is constructed by means of catenary arcs leaning on the walls, whose endpoints change in depth in order to fit the structure. We approximate its form by fixing the maximum height of the roof at $r_2 = 0.7$ and the space curve

$$\alpha_3(u) = (u, -\frac{w}{2}, \frac{r_2}{2}\cos(\omega_1 u) + h + \frac{r_2}{2}), \quad 0 < u < \ell_1$$

as the maximum of the catenary arc. Therefore, for every $0 < u < \ell_1$, the catenary arc passes through the points $\alpha_1(u)$, $\alpha_2(u)$, and $\alpha_3(u)$, with $\alpha_3(u)$ being its maximum.

The parametrization determining a catenary arc in each plane is given by

$$\left(u, v, \frac{a(u)}{2}(e^{\frac{v}{a(u)}} + e^{-\frac{v}{a(u)}} + b(u)), \quad r\cos(\omega_1 u - \omega_2)\right) - w < v < r\cos(\omega_1 u),$$

Fig. 4.18 Scheme for a catenary arc in the vault of the church of Cristo Obrero

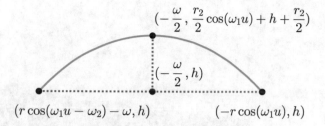

$$\left(-\frac{\omega}{2}, \frac{r_2}{2}\cos(\omega_1 u) + h + \frac{r_2}{2}\right)$$

$$\left(-\frac{\omega}{2}, h\right)$$

$$(r\cos(\omega_1 u - \omega_2) - \omega, h) \qquad\qquad (-r\cos(\omega_1 u), h)$$

for some $a(u), b(u) \in \mathbb{R}$, which depend on the depth parameter u (see Sect. 1.2). The value $a(u)$ is the parameter in the definition of the catenary arc, whereas $b(u)$ moves the catenary vertically. Its symmetry point is located at $v = \omega/2$, so the catenary arc is parametrized after this shifting by

$$\left(u, v, \frac{a(u)}{2}\left(e^{\frac{v+\omega/2}{a(u)}} + e^{-\frac{v+\omega/2}{a(u)}}\right) + b(u)\right), \quad -r\cos(\omega_1 u) - w < v < r\cos(\omega_1 u)).$$

The situation for every value of u is shown in Fig. 4.18.

The values of $a(u)$ and $b(u)$ are obtained by approximating the solution of the system when imposing the interpolating conditions, i.e.,

$$\begin{cases} \frac{a(u)}{2}\left(\exp(\frac{r\cos(\omega_1 u)+\omega/2}{a}) + \exp(-\frac{r\cos(\omega_1 u)+\omega/2}{a})\right) + b(u) = h \\ a(u) + b(u) = r_2 \end{cases}$$

We approximate by means of a Taylor expansion the elements in the exponentials involved and consider

$$a(u) \approx \frac{(r\cos(\omega_1 u) + \frac{\omega}{2})^2}{2(h - r_2)}, \quad b(u) \approx r_2 - a(u).$$

The vault is then parametrized by

$$\left(u, v, \frac{a(u)}{2}\left(\exp(\frac{v + \omega/2}{a(u)}) + \exp(-\frac{v + \omega/2}{a(u)})\right) + b(u)\right),$$

for the previous values of $a(u)$ and $b(u)$, $0 < u < \ell_1$ and $-r\cos(\omega_1 u) - \omega < v < r\cos(\omega_1 u)$.

Figure 4.19 shows the configuration of the vault studied and the limiting curves. Both the walls and the vault represented in Fig. 4.19, with the help of Maple software (Fig. 4.20).

Other interpolation techniques applied nowadays in architectural studies can be found in Bärtschi et al. (2010).

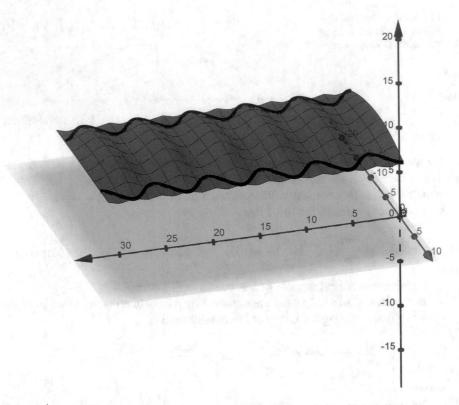

Fig. 4.19 Scheme for the vault of the church of Cristo Obrero

4.4 Surfaces of Revolution

A **surface of revolution** is a surface determined by the rotation of a space curve, known as the **generatrix** around a line which is the **axis of rotation**.

Worthy of mention is the following classic result, known as Rodrigues' rotation formula.

Theorem 4.4.1 (Rodrigues' Rotation Formula) *Let r be a line in $\mathbb{R}^3$ with direction $\vec{n} = (n_1, n_2, n_3) \in \mathbb{R}^3 \setminus \{(0, 0, 0)\}$ and let us assume that the origin of coordinates belongs to r, and that $\|\vec{n}\| = 1$. Let $\theta \in [0, 2\pi)$. It holds that the coordinates of the rotation of a point $P = (P_1, P_2, P_3) \in \mathbb{R}^3$ around the line r and angle θ is determined by*

$$\cos(\theta)\vec{OP} + (1 - \cos(\theta))(\vec{OP} \cdot \vec{n})\vec{n} + \sin(\theta)(\vec{n} \times \vec{OP}). \tag{4.7}$$

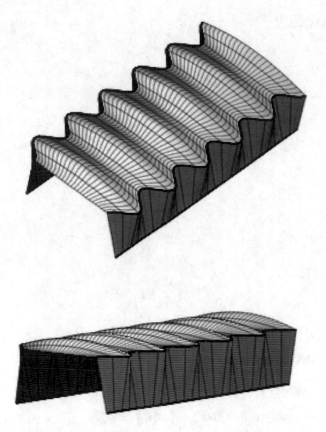

Fig. 4.20   Scheme of the church of Cristo Obrero performed in Maple

Proof Equation (4.7) is a consequence of the following geometric reasoning. Figure 4.21 illustrates the situation. We assume that P does not belong to r. Otherwise, (4.7) can be checked by direct computations.

Let O' be the projection of P (or Q) on the line r. We have $O' \neq P$, and $O' \neq Q$. We consider the orthonormal basis of $\mathbb{R}^3$

$$\left\{ \frac{\vec{O'P}}{\left\| \vec{O'P} \right\|}, \frac{\vec{n} \times \vec{O'P}}{\left\| \vec{n} \times \vec{O'P} \right\|}, \vec{n} \right\} =: \{\vec{e}_1, \vec{e}_2, \vec{n}\}.$$

In this basis, we have

$$\vec{O'Q} = \left\| \vec{O'Q} \right\| \cos(\theta)\vec{e}_1 + \left\| \vec{O'Q} \right\| \sin(\theta)\vec{e}_2.$$

Fig. 4.21 Rodrigues'
rotation formula

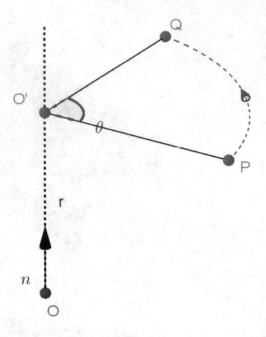

By direct computations, we derive that the previous expression equals

$$\cos(\theta)\vec{O'P} + \|O'P\| \sin(\theta)\vec{e}_2$$
$$= \cos(\theta)(\vec{OP} - \vec{OO'}) + \|O'P\| \sin(\theta)\vec{e}_2$$
$$= \cos(\theta)(\vec{OP} - (\vec{OP} \cdot \vec{n})\vec{n}) + \|O'P\| \sin(\theta)\vec{e}_2$$
$$= \cos(\theta)(\vec{OP} - (\vec{OP} \cdot \vec{n})\vec{n}) + \sin(\theta)(\vec{n} \times (\vec{OP} - \vec{OO'}))$$
$$= \cos(\theta)(\vec{OP} - (\vec{OP} \cdot \vec{n})\vec{n}) + \sin(\theta)(\vec{n} \times \vec{OP}).$$

The conclusion follows from the fact that $\vec{OO'} + \vec{O'Q} = \vec{OQ}$ in the form

$$\vec{O'Q} = \vec{OQ} - (\vec{OP} \cdot \vec{n})\vec{n}.$$

One can rewrite Rodrigues' formula in matrix form as follows.

Corollary 4.4.2 *Under the same assumptions of Theorem 4.4.1, we have that given
a point $P = (x, y, z) \in \mathbb{R}^3$, then the rotation of P around the line of direction
$\vec{n} = (n_1, n_2, n_3)$ is the point $Q = (x', y', z')$ of angle $\theta \in \mathbb{R}$, which satisfies*

$$\begin{pmatrix} x' \\ y' \\ z' \end{pmatrix} = A(\theta) \begin{pmatrix} x \\ y \\ z \end{pmatrix}, \tag{4.8}$$

where $A(\theta)$ is the matrix

$$\begin{pmatrix} \cos(\theta) + n_1^2(1 - \cos(\theta)) & n_1 n_2(1 - \cos(\theta)) - n_3 \sin(\theta) & n_1 n_3(1 - \cos(\theta)) + n_2 \sin(\theta) \\ n_1 n_2(1 - \cos(\theta)) + n_3 \sin(\theta) & \cos(\theta) + n_2^2(1 - \cos(\theta)) & n_2 n_3(1 - \cos(\theta)) - n_1 \sin(\theta) \\ n_1 n_3(1 - \cos(\theta)) - n_2 \sin(\theta) & n_2 n_3(1 - \cos(\theta)) + n_1 \sin(\theta) & \cos(\theta) + n_3^2(1 - \cos(\theta)) \end{pmatrix}.$$

Observe that the rotation of a point around the coordinate axis OX, OY and OZ, of angle θ, is represented by the matrices

$$\begin{pmatrix} 1 & 0 & 0 \\ 0 & \cos(\theta) & -\sin(\theta) \\ 0 & \sin(\theta) & \cos(\theta) \end{pmatrix}, \quad \begin{pmatrix} \cos(\theta) & 0 & \sin(\theta) \\ 0 & 1 & 0 \\ -\sin(\theta) & 0 & \cos(\theta) \end{pmatrix}, \quad \begin{pmatrix} \cos(\theta) & -\sin(\theta) & 0 \\ \sin(\theta) & \cos(\theta) & 0 \\ 0 & 0 & 1 \end{pmatrix},$$

respectively.

We now consider a line r not necessarily passing through the origin of coordinates. Let $\vec{n} = (n_1, n_2, n_3)$ be a unitary vector with the direction of the line, and let $P_0 = (x_0, y_0, z_0)$ be a point of the line. We want to compute the rotation of P around r of angle θ. It is sufficient to proceed using the following algorithm.

Algorithm

Input: r, P_0, θ.

Output: The rotated point Q_0, obtained from rotating P_0 around r an angle θ.

1. Compute the translation of vector $-\vec{OP_0}$. Let r' be the translated line and P_0' the translated point.
2. Compute the rotation of P_0' around r' of angle θ with Rodrigues' formula. Let Q_0' be the resulting point.
3. Let $Q_0 = Q_0' + \vec{OP_0}$ be the translation of Q_0' after the inverse translation.
4. Return Q_0.

The above construction allows us to determine surfaces of revolution by storing all the rotated points in a parametrized curve within the action of two parameters: one of them drawing the curve, and the other constructing the corresponding circles in the surface of revolution.

In this regard, one has the following construction of a surface of revolution associated to a parametrized space curve. Let r be a line passing through the origin and with $\vec{n} = (n_1, n_2, n_3)$ being a unitary vector in the direction of the line, and let (I, α) be a parametrized space curve, with $\alpha(t) = (\alpha_1(t), \alpha_2(t), \alpha_3(t))$, for all $t \in I$. The revolution surface associated to (I, α) and r is parametrized by $X : I \times \mathbb{R} \to \mathbb{R}^3$, with $X(u, v) = (x(u, v), y(u, v), z(u, v))$ and

$$\begin{pmatrix} x(u, v) \\ y(u, v) \\ z(u, v) \end{pmatrix} = A(v) \begin{pmatrix} \alpha_1(u) \\ \alpha_2(u) \\ \alpha_3(u) \end{pmatrix},$$

where the matrix $A(v)$ is defined in (4.8).

Example 4.4.3 In this example, we construct the surface of revolution consisting
of the revolution of a line (generatrix) with respect to an axis of rotation such that
both are skewed lines, i.e., non-parallel lines which do not intersect. For the sake
of simplicity, we consider the axis of rotation to be the coordinate axis OX, and
$\vec{n} = (0, 0, 1)$. The generatrix is the line joining the points $P_1 = (1, 0, -1)$ and
$P_2 = (\cos(s), \sin(s), 1)$, for some $s \in (0, \pi)$.

As mentioned, the generatrix and the axis of rotation are skew lines. Let

$$(\alpha_1(t), \alpha_2(t), \alpha_3(t)) = (1, 0, -1) + t(\cos(s) - 1, \sin(s), 2), \quad t \in \mathbb{R}$$

be a parametrization of the generatrix. Therefore, the associated revolution surface can be parametrized by (U, X), where $U = \mathbb{R} \times \mathbb{R}$ and $X(u, v) =$
$(x(u, v), y(u, v), z(u, v))$, with

$$\begin{pmatrix} x(u, v) \\ y(u, v) \\ z(u, v) \end{pmatrix} = A(v) \begin{pmatrix} \alpha_1(u) \\ \alpha_2(u) \\ \alpha_3(u) \end{pmatrix}$$

determined in this specific case by

$$\begin{cases} x(u, v) = \cos(v)(1 + u(\cos(s) - 1)) - \sin(v)u\sin(s) \\ y(u, v) = \sin(v)(1 + u(\cos(s) - 1)) + \cos(v)u\sin(s) \\ z(u, v) = 2u - 1, \end{cases}$$

for $(u, v) \in \mathbb{R} \times \mathbb{R}$. In order to provide with an implicit equation of this
surface, we solve the system determined by the first two elements in the previous
parametrizations in $\cos(v)$ and $\sin(v)$. This yields

$$\cos(v) = \frac{\cos(s)ux + \sin(s)uy - ux + x}{\cos(s)^2u^2 + \sin(s)^2u^2 - 2\cos(s)u^2 + 2\cos(s)u + u^2 - 2u + 1},$$

$$\sin(v) = \frac{\cos(s)uy - \sin(s)ux - uy + y}{\cos(s)^2u^2 + \sin(s)^2u^2 - 2\cos(s)u^2 + 2\cos(s)u + u^2 - 2u + 1}.$$

Taking into account the trigonometric equality $\cos(v)^2 + \sin(v)^2 = 1$ we derive

$$x^2 + y^2 = -2\cos(s)u^2 + 2\cos(s)u + 2u^2 - 2u + 1.$$

Finally, the parametrization of the surface in the third coordinate $z(u, v) = 2u - 1$
yields

$$x^2 + y^2 = -2\cos(s)\left(\frac{z+1}{2}\right)^2 + 2\cos(s)\frac{z+1}{2} + 2\left(\frac{z+1}{2}\right)^2 - 2\frac{z+1}{2} + 1,$$

which is equivalent to

$$2x^2 + 2y^2 + (\cos(s) - 1)z^2 - \cos(s) - 1 = 0.$$

This last expression determines the equation of a surface for every choice of the parameter s. From the construction of the surface, we can guarantee that it is a surface of revolution so that certain symmetries with respect to the OZ axis can be stated. In addition to that, it is a ruled surface, and the function of two variables determining its implicit equation is a polynomial in three variables. This type of surface is known as a quadric. The theory described in Sect. 4.5 will state that this surface in fact represents a hyperboloid of one sheet for any choice of the parameter $s \in (0, \pi)$.

It is worth mentioning that this surface of revolution is in fact a doubly ruled surface, i.e., two families of lines can describe the surface. It is not difficult to verify that the line

$$(\beta_1(t), \beta_2(t), \beta_3(t)) = (1, 0, -1) + t(\cos(s) - 1, -\sin(s), 2)$$

is contained in it. We leave it up to the reader to verify that the same rotation as above describes the same surface. This double ruling provides stability properties to the structure, which is widely used in architecture (Fig. 4.22).

One can also consider a plane curve rotating around an axis. In this particular case, the formulation is simplified. Assume that

$$\alpha(t) = (\alpha_1(t), 0, \alpha_3(t)), \quad t \in I,$$

and the axis of revolution is the coordinate axis OZ, i.e., $\vec{n} = (0, 0, 1)$. Then the surface of revolution can be parametrized by (U, X), with $U = I \times \mathbb{R}$ and $X : U \to \mathbb{R}^3$ and

$$X(u, v) = (\alpha_1(u)\cos(v), \alpha_1(u)\sin(v), \alpha_3(u)),$$

for every $u \in I$ and $v \in \mathbb{R}$.

Observe that in plan view a circle is determined for every fixed $u \in I$. For every choice of v, the curve turns out to be the rotation of angle v of the initial curve with OZ being the axis of revolution. In order to illustrate this situation, we consider a catenary-like curve parametrized by

$$\alpha(t) = (t, a\cosh\left(\frac{t}{a}\right) + 5), \quad t \in (-a\,\text{arccosh}(-5/a), a\,\text{arccosh}(-5/a)),$$

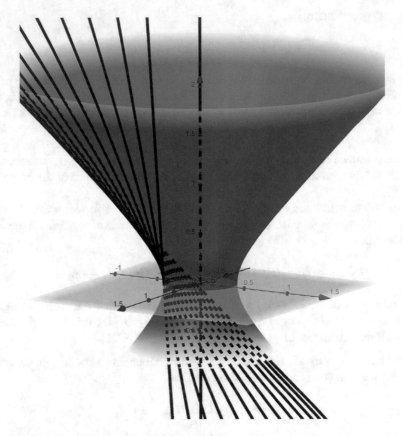

Fig. 4.22 The surface of revolution in Example 4.4.3

for $a < 0$. The surface of revolution described by this curve rotating around the OZ axis is parametrized by

$$X(u, v) = (u \cos(v), u \sin(v), a \cosh\left(\frac{u}{a}\right) + 5),$$

for $u \in (0, 2\pi)$ and $v \in (-a\operatorname{arccosh}(-5/a), a\operatorname{arccosh}(-5/a))$. The importance of the forces acting on this structure are due to the catenary arcs involved. Figure 4.23 shows this surface of revolution for $a = -5/2$, which resembles an igloo.

Let us consider the circle of radius $r > 0$, located in the plane $x = 0$, and centered at the point $(0, R, 0)$, for $r < R$. We choose the OZ axis as the axis of revolution, and obtain a **torus**, parametrized by

$$X(u, v) = (-\sin(u)(r \cos(v) + R), \cos(u)(r \cos(v) + R), r \sin(v)),$$

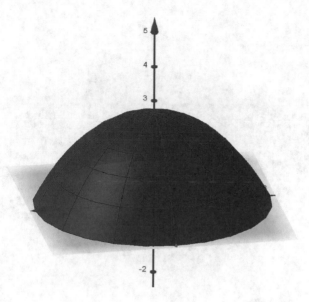

Fig. 4.23 Surface of revolution

with $u, v \in (0, 2\pi)$. This comes as a result of considering the parametrization of the circle $(0, r\cos(v) + R, r\sin(v))$, for $v \in (0, 2\pi)$, and the parametrization derived from Rodrigues' formula

$$\begin{pmatrix} x(u, v) \\ y(u, v) \\ z(u, v) \end{pmatrix} = \begin{pmatrix} \cos(u) & -\sin(u) & 0 \\ \sin(u) & \cos(u) & 0 \\ 0 & 0 & 1 \end{pmatrix} \begin{pmatrix} 0 \\ r\cos(v) + R \\ r\sin(v) \end{pmatrix}.$$

Figure 4.24 illustrates a specific torus for $r = 1$ and $R = 3$.

Such structures appear or inspire architectural elements, for example in the Comptoir forestier by Samyn and Partners, mentioned earlier. In Narváez-Rodríguez and Barrera-Vera (2016), Roberto Narváez-Rodríguez and José Antonio Barrera-Vera work on the conical components for rotational parabolic domes.

Other generalizations of surfaces of revolution such as the torus are channel (or canal) surfaces, whose parametrization has been studied in different works such as Peternell and Pottmann (1997), Landsmann et al. (2001). Those surfaces generalize the torus surfaces of revolution in the sense that they are formed with the envelope of the spheres whose centers lie on a space curve.

Example 4.4.4 In Sect. 4.4, we have gone into detail about the surfaces obtained by the revolution of some curve with respect to an axis. Here, we apply the theory to a specific building.

Let us consider the thin-shell water tower in Fedala, designed by Eduardo Torroja. Its main structure can be approximated by the surface of revolution of two

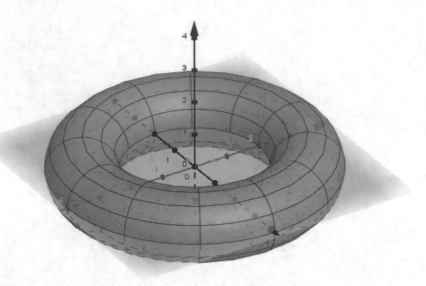

Fig. 4.24 Torus; $r = 1$, $R = 3$

hyperbolas which join at a point. Let us assume that the hyperbolas are drawn in the XZ plane, and that the revolution axis is the line $\{x = y = 0\}$.

We consider a scaled version of the building, so that, regarding the different views, the two hyperbolas are determined by the space curves (I_1, α_1) and (I_2, α_2), where

$$\alpha_1(t) = (-\sqrt{a_1 t^2 + b_1} + c_1, 0, t - \alpha_2), \quad I_1 = (\alpha_1, \beta_1)$$

and

$$\alpha_2(t) = (-\sqrt{a_2 t^2 + b_2} + c_2, 0, t - \alpha_2), \quad I_1 = (\alpha_2, \beta_2),$$

where $\alpha_1 = 0$, $\beta_1 = 1.4$, $\alpha_2 = -1.14$, $\beta_2 = 0$; $a_1 = 2.83$, $b_1 = 1.7$, $c_1 = 0.22$, $a_2 = 1.35$, $b_2 = 1.47$, $c_2 = 0.135$. These coefficients have been obtained by inspection. Following Sect. 4.4, one constructs both surfaces of revolution as the surfaces which have the parametrizations (U_1, X_1) and (U_2, X_2) for $U_1 = (\alpha_1, \beta_1) \times (0, 2\pi)$ and $U_2 = (\alpha_2, \beta_2) \times (0, 2\pi)$, and with

$$X_1(u, v) = (\cos(v)\alpha_1(u), \sin(v)\alpha_1(u), u),$$

$$X_2(u, v) = (\cos(v)\alpha_2(u), \sin(v)\alpha_2(u), u).$$

The result is displayed in Fig. 4.25.

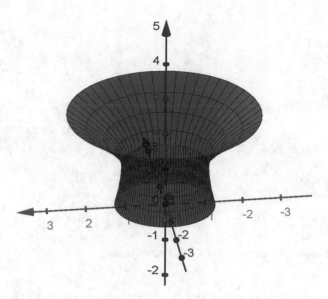

Fig. 4.25 Approximation of the Water tower in Fedala as a surface of revolution

4.5 Quadric Surfaces

The structure of this section is planned to follow that of Sect. 1.4 on conic curves. There we considered those plane curves implicitly defined by the zeros of a polynomial in two variables. Here we deal with zeros of polynomials of second degree in three variables which lead to the implicit description of a surface known as a quadric.

As a consequence, a **quadric surface** is defined by the set

$$S = \{(x, y, z) \in \mathbb{R}^3 : P(x, y, z) = 0\}, \tag{4.9}$$

for some second degree polynomial P, i.e.,

$$P(x, y, z) = a_{11}x^2 + a_{22}y^2 + a_{33}z^2 + 2a_{12}xy + 2a_{13}xz$$
$$+ 2a_{23}yz + 2a_{01}x + 2a_{02}y + 2a_{03}z + a_{00},$$

for some $a_{ij} \in \mathbb{R}, 0 \le i, j \le 3$. A point $\boldsymbol{x} = (x, y, z) \in \mathbb{R}^3$ belongs to the surface S in (4.9) if it holds that

$$\begin{pmatrix} 1 & x & y & z \end{pmatrix} \begin{pmatrix} a_{00} & a_{01} & a_{02} & a_{03} \\ a_{01} & a_{11} & a_{12} & a_{13} \\ a_{02} & a_{12} & a_{22} & a_{23} \\ a_{03} & a_{13} & a_{23} & a_{33} \end{pmatrix} \begin{pmatrix} 1 \\ x \\ y \\ z \end{pmatrix} = \boldsymbol{x}^T M \boldsymbol{x} = 0.$$

Fig. 4.26 Orthogonal transformation of a coordinate system

At this point, we proceed in a way that is analogous to what we did in Chap. 1. Using **orthogonal transformations** in space, i.e., endomorphisms from the vector space $\mathbb{R}^3$ into itself which preserve the scalar product:

$$\langle v_1, v_2 \rangle = \langle f(v_1), f(v_2) \rangle, \quad \text{for all } v_1, v_2 \in \mathbb{R}^3$$

(see Definition 1.4.1 for the two-dimensional version of the definition of an orthogonal transformation). Let there be two affine bases of $\mathbb{R}^3$, say $\{O, \{e_1, e_2, e_3\}\}$ and $\{O', \{u_1, u_2, u_3\}\}$. Here $O, O' \in \mathbb{R}^3$ play the role of the origin of coordinates, and $\{e_1, e_2, e_3\}$ and $\{u_1, u_2, u_3\}$ are orthogonal bases of the associated vector space. It holds that one coordinate system can be transformed into the other by means of an orthogonal transformation (isometry) (see Fig. 4.26).

These types of transformations will allow us to classify any quadric, since the essence of the procedure is analogous to that described for conics in Sect. 1.4 (see also de Burgos Román (2006)). Therefore, any matrix representing a quadric M and its representation M' in a new orthogonal coordinate system are related by means of a change of affine coordinates

$$Q = \left(\begin{array}{c|c} 1 & O \\ \hline c & Q_0 \end{array} \right),$$

where $c^T = (c_1, c_2, c_3)$ turns out to be the coordinates of the first origin of affine coordinates in the new reference, and Q_0 is the orthogonal matrix associated to the change of bases in the associated vector space, $\mathbb{R}^3$. Let us write

$$M = \left(\begin{array}{c|c} d & b^T \\ \hline b & M_0 \end{array} \right),$$

with $d \in \mathbb{R}$, and $b \in \mathcal{M}_{3 \times 1}(\mathbb{R})$. The spectral theorem allows us to choose Q_0 such that $M_0' = Q_0^T M_0 Q_0$ is a diagonal matrix in the form $M_0' = \mathrm{diag}(\lambda_1, \lambda_2, \lambda_3)$ for some $\lambda_1, \lambda_2, \lambda_3 \in \mathbb{R}$. This matrix is constructed by means of the corresponding eigenvalues of M_0. We refer to Strang (1993) for further details in this direction. Different elements of the quadric remain unchanged after the change of coordinates, and will allow us to formulate a classification of all the quadric surfaces. These elements are related to the coefficients or the zeros of the characteristic polynomial associated to M_0, which coincides with that of M_0', as they are similar matrices. Among these invariant elements we have

- the determinant of M_0, given by $\lambda_1 \lambda_2 \lambda_3$,
- the signature of M_0, given by $\lambda_1 + \lambda_2 + \lambda_3$,
- the value of $\lambda_1 \lambda_2 + \lambda_1 \lambda_3 + \lambda_2 \lambda_3$,
- the determinant of M,
- the rank of M,
- etc.

More precisely, observe that the coefficient of the quadratic term of the characteristic polynomial of M_0 (or equivalently M_0') is (apart from a sign) the signature of M_0, and the coefficient of the linear term is the third element in the previous list.

Assume we have made a change of coordinates as before, leading to a diagonal matrix M_0'. Then it holds that the transformed matrix of the quadric is given by

$$
\begin{pmatrix}
\delta & \alpha & \beta & \gamma \\
\alpha & \lambda_1 & 0 & 0 \\
\beta & 0 & \lambda_2 & 0 \\
\gamma & 0 & 0 & \lambda_3
\end{pmatrix},
\tag{4.10}
$$

for some $\alpha, \beta, \gamma, \delta \in \mathbb{R}$.

In what follows, we come up with all the different quadrics regarding the values of the invariant elements. We distinguish the following cases to be studied:

Case 1. $\lambda_1 \lambda_2 \lambda_3 \neq 0$
Case 2. $\lambda_1 \lambda_2 \lambda_3 = 0$.

In the following several subdivisions appear for each case that are denoted in accordance with the previous notation.

First, let us assume that $\lambda_1 \lambda_2 \lambda_3 \neq 0$. Then we can carry out the transformation

$$
(x, y, z) \mapsto (x - \alpha/\lambda_1, y - \beta/\lambda_2, z - \gamma/\lambda_3),
$$

which turns out to be a translation, and the matrix of the quadric after such transformation becomes

$$
\begin{pmatrix}
k & 0 & 0 & 0 \\
0 & \lambda_1 & 0 & 0 \\
0 & 0 & \lambda_2 & 0 \\
0 & 0 & 0 & \lambda_3
\end{pmatrix},
$$

for some $k \in \mathbb{R}$. The invariant elements of the quadric surfaces guarantee that

$$k = \frac{\det(M)}{\det(M_0)},$$

which yields the equation

$$\lambda_1 (x^{\star\star})^2 + \lambda_2 (y^{\star\star}) + \lambda_3 (z^{\star\star}) = -\frac{\det(M)}{\det(M_0)} \qquad (4.11)$$

representing the quadric in the new coordinate system. The resulting quadric is said to be in canonical form. At this point, we distinguish two subcases:

1.1 $\det(M) \neq 0$
1.2 $\det(M) = 0$

In Case 1.1, it is possible to figure out which quadric we are starting from:

1.1.1 If $\lambda_1, \lambda_2, \lambda_3$ and $-\frac{\det(M)}{\det(M_0)}$ are of the same sign, then the quadric is an **ellipsoid** (Fig. 4.27).

1.1.2 If $\lambda_1, \lambda_2, \lambda_3$ are of the same sign, and and $-\frac{\det(M)}{\det(M_0)}$ has different sign, then no point can satisfy (4.11), and the quadric is empty. It is known as an **imaginary ellipsoid**, by similarity to the previous situation.

1.1.3 If two of the three numbers λ_1, λ_2 and λ_3 share sign with $-\frac{\det(M)}{\det(M_0)}$, the quadric is a **hyperbolid of one sheet**.

1.1.4 If two of the three numbers λ_1, λ_2 and λ_3 are of the same sign, and differ from the sign of $-\frac{\det(M)}{\det(M_0)}$, then the quadric is a **hyperboloid of two sheets**.

Let us say a few words about the above quadric surfaces.

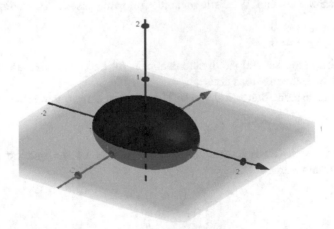

Fig. 4.27 Ellipsoid

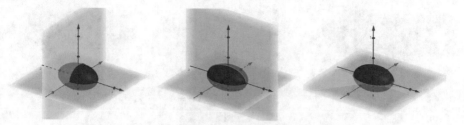

Fig. 4.28 Sections of an ellipsoid in canonical form by the coordinate planes

Fig. 4.29 Hyperboloid of one sheet

The canonical form of an ellipsoid is given by

$$(x^{\star\star})^2/a^2 + (y^{\star\star})^2/b^2 + (z^{\star\star})^2/c^2 = 1.$$

Observe that its sections by the coordinate planes determine ellipses of semiaxis a, b or c depending on the plane chosen for the section (Fig. 4.28).

The canonical form of a hyperboloid of one sheet (Fig. 4.29) is given by

$$(x^{\star\star})^2/a^2 + (y^{\star\star})^2/b^2 - (z^{\star\star})^2/c^2 = 1.$$

Concerning this quadric, we can observe that the sections by the coordinate planes describe ellipses or hyperbolas, depending on the plane. This can be checked by substituting each of the variables by 0 in the equation of its canonical form (Fig. 4.30).

Fig. 4.30 Sections of a hyperboloid of one sheet in canonical form by the coordinate planes

Example 4.5.1 Let us reconsider Example 4.4.3. We have observed that the surface of revolution obtained by the rotation of a line in skew position with respect to the axis of revolution determines a hyperboloid of one sheet. More precisely, let us consider the line parametrized by

$$(\alpha_1(t), \alpha_2(t), \alpha_3(t)) = (1, 0, -1) + t(\cos(s) - 1, \sin(s), 2), \quad t \in \mathbb{R},$$

where $s \in \mathbb{R}$ is a fixed number. The axis of revolution is the OZ coordinate axis. The points of the revolution surface satisfy the equation

$$2x^2 + 2y^2 + \cos(s)z^2 - \cos(s) - z^2 - 1 = 0.$$

We observe that the matrix associated to this quadric is

$$M = \begin{pmatrix} -1 - \cos(s) & 0 & 0 & 0 \\ 0 & 2 & 0 & 0 \\ 0 & 0 & 2 & 0 \\ 0 & 0 & 0 & \cos(s) - 1 \end{pmatrix}$$

Therefore, $\lambda_1 = \lambda_2 = 2$ and $\lambda_3 = \cos(s) - 1$. The previous statements allow us to conclude that the quadric is a hyperboloid of one sheet for every $s \neq k\pi$, for $k \in \mathbb{Z}$. If $s = 2k\pi$ for some $k \in \mathbb{Z}$, the quadric is of equation $x^2 + y^2 = 1$, i.e., it represents a cylinder. In the case that $s = (2k+1)\pi$ for some $k \in \mathbb{Z}$, the quadric is of equation $x^2 + y^2 = z^2$, which represents a cone.

On the other hand, the canonical form of a hyperboloid of two sheets (Fig. 4.31) is

$$(x^{\star\star})^2/a^2 + (y^{\star\star})^2/b^2 - (z^{\star\star})^2/c^2 = -1.$$

The sections of a hyperboloid of two sheets in canonical form by the coordinate planes lead to a hyperbola (for $y^{\star\star} = 0$ or $x^{\star\star} = 0$), or an empty intersection (for $z^{\star\star} = 0$) (Fig. 4.32).

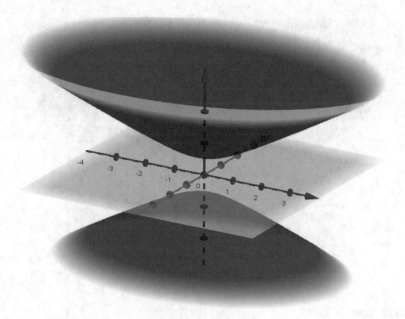

Fig. 4.31 Hyperboloid of two sheets

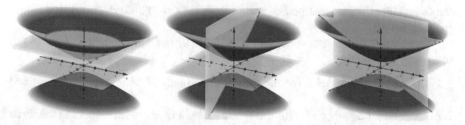

Fig. 4.32 Sections of a hyperboloid of two sheets in canonical form by the coordinate planes

We now pay attention to the situation described in the case 1.2 distinguished above, i.e., $\det(M) = 0$. The canonical form of the quadric is given by

$$\lambda_1(x^{\star\star})^2 + \lambda_2(y^{\star\star})^2 + \lambda_3(z^{\star\star})^2 = 0.$$

We distinguish two situations:

1.2.1 λ_1, λ_2 and λ_3 do not share their sign. We obtain a **cone** (Fig. 4.33).
1.2.2 λ_1, λ_2 and λ_3 share their sign. The surface is said to be an **imaginary cone**.

The notation for imaginary cone is taken from that for the cone, by imitation, although the set of points in an imaginary cone is reduced to a point, the vertex of the cone. This point is the origin of coordinates when working with the cone in canonical form. With respect to the cone in canonical form, it holds that its cuts with

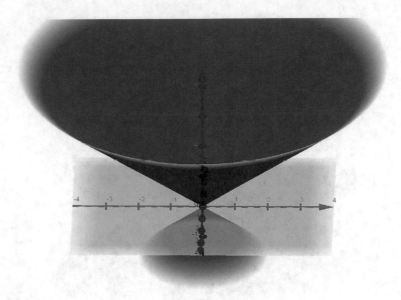

Fig. 4.33 Cone

the coordinate planes are a pair of secant lines ($x^{\star\star} = 0$ and $y^{\star\star} = 0$) or a single point ($z^{\star\star} = 0$).

We proceed with the case 2., i.e. $\lambda_1\lambda_2\lambda_3 = 0$, and split it with regard to the following statements:

2.1 $\lambda_1\lambda_2 \neq 0$ and $\lambda_3 = 0$ (or the symmetric situations with respect to the three variables)

2.2 $\lambda_1 \neq 0$ and $\lambda_2 = \lambda_3 = 0$ (or the symmetric situations with respect to the three variables)

2.3 $\lambda_1 = \lambda_2 = \lambda_3 = 0$.

The last situation leads to a **double plane**, as all the quadratic terms vanish in the canonic form of the quadric. The two other items require deeper study.

First, assume that $\lambda_1\lambda_2 \neq 0$ and $\lambda_3 = 0$. After writing the quadric surface in the form (4.10), we can perform the translation

$$(x, y, z) \mapsto (x - \frac{\alpha}{\lambda_1}, y - \frac{\beta}{\lambda_2}, z).$$

This transformation causes the matrix associated to the quadric to turn into

$$\begin{pmatrix} k_1 & 0 & 0 & k_2 \\ 0 & \lambda_1 & 0 & 0 \\ 0 & 0 & \lambda_2 & 0 \\ k_2 & 0 & 0 & 0 \end{pmatrix},$$

for some $k_1, k_2 \in \mathbb{R}$, which can be determined from the invariant elements of the matrix of a given quadric surface, namely one observes that $k_2^2 = \frac{\det(M)}{\lambda_1 \lambda_2}$, and some $k_1 \in \mathbb{R}$. That k_1 is irrelevant at this point because the second transformation on the quadric

$$(x, y, z) \mapsto (x, y, z - \frac{k_1}{2k_2})$$

turns it into the form

$$\lambda_1(x^{\star\star})^2 + \lambda_2(y^{\star\star})^2 + 2k_2 z = 0, \quad k_2 = \pm\sqrt{\frac{\det(M)}{\lambda_1 \lambda_2}}.$$

We distinguish the two following cases:

2.1.1 $k_2 \neq 0$
2.1.2 $k_2 = 0$

In Case 2.1.1, we also make the following distinction:

2.1.1.1 $\lambda_1 \lambda_2 > 0$, obtaining an **elliptic paraboloid** (Fig. 4.34).
2.1.1.2 $\lambda_1 \lambda_2 < 0$, which leads to a **hyperbolic paraboloid**.

The sections of an elliptic paraboloid in canonical form by the coordinate planes is a parabola ($x^{\star\star} = 0$ or $y^{\star\star} = 0$) or the point $O = (0, 0, 0)$ ($z^{\star\star} = 0$) (Fig. 4.35).

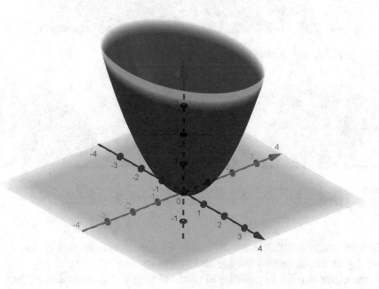

Fig. 4.34 Elliptic paraboloid

Fig. 4.35 Sections of an elliptic paraboloid in canonical form at positive height (left) and with $y = 0$ (right)

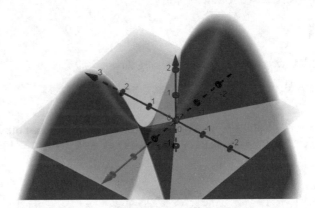

Fig. 4.36 Hyperbolic paraboloid

Fig. 4.37 Sections of a hyperbolic paraboloid in canonical form at positive (left), negative (center) and null (right) height

The sections at positive or negative height either describe an ellipse or are empty, depending on the sign of λ_1 (or λ_2) and that of k_2.

On the other hand, a hyperbolic paraboloid (Fig. 4.36) describes a pair of secant lines at the floor, and hyperbolas at other heights, whereas the sections by the other coordinate planes produce parabolas (Figs. 4.37, 4.38).

Fig. 4.38 Sections of a hyperbolic paraboloid in canonical form with the planes $x = 0$ and $y = 0$

The following stage to be studied is Case 2.1.2. Here, $k_2 = 0$ and the equation of the quadric becomes

$$\lambda_1 x^2 + \lambda_2 y^2 + k_1 = 0,$$

for $k_1 \in \mathbb{R}$, which can be determined in terms of the invariants of the quadric after the rigid movements described. We distinguish several situations depending on the sign of the elements involved.

2.1.2.1 If $k_1 \neq 0$ and λ_1, λ_2 and k_1 are of the same sign, the quadric is empty, and by proximity to the next case, we call this quadric surface the **imaginary cylinder**.

2.1.2.2 If $k_1 \neq 0$ and $\lambda_1 \lambda_2 > 0$ but the sign of k_1 differs from that of λ_1 (or λ_2), the quadric is an **elliptic cylinder**, i.e., a cylinder whose directrix is an ellipse.

2.1.2.3 If $k_1 \neq 0$ and $\lambda_1 k_1 > 0$ but the sign of λ_2 differs from that of λ_1 (or k_1), or the symmetric situation with respect to the eigenvalues, the quadric is a **hyperbolic cylinder**, i.e., a cylinder based on a hyperbola.

2.1.2.4 If $k_1 = 0$, the reduced equation of the quadric surface is $\lambda_1 x^2 + \lambda_2 y^2 = 0$. If $\lambda_1 \lambda_2 < 0$, the equation can be factorized and the quadric turns out to be a pair of **secant planes**.

2.1.2.5 Otherwise, if $\lambda_1 \lambda_2 > 0$ and still $k_1 = 0$, the quadric is empty, resulting in a pair of secant complex planes.

We stress at this point that the value of k_1 (which we do not determine explicitly) can be obtained after the transformations and, in order to classify the quadric, this could be done via invariants, or even by different sections of the quadric by a plane.

We conclude with the analysis of Case 2.2, i.e., if $\lambda_1 \neq 0, \lambda_2 = \lambda_3 = 0$. Here the translation

$$(x, y, z) \mapsto \left(x - \frac{\alpha}{\lambda_1}, y, z\right)$$

can be performed, arriving at the equation

$$\lambda_1 (x^{\star\star})^2 + 2\tilde{k}_3 y^{\star\star} + 2\tilde{k}_2 z^{\star\star} + \tilde{k}_1 = 0,$$

for some $\tilde{k}_1, \tilde{k}_2, \tilde{k}_3 \in \mathbb{R}$ which can be obtained from the invariants of the quadric surface. Indeed, $\tilde{k}_1 = \delta - \alpha^2/\lambda_1$, $\tilde{k}_2 = \gamma$ and $\tilde{k}_3 = \beta$. In this situation, the classification is made by taking into account the rank of M, or equivalently, the nullity of $(\tilde{k}_2, \tilde{k}_3)$.

2.2.1 If $(\tilde{k}_2, \tilde{k}_3) \neq (0, 0)$, the quadric is a **parabolic cylinder**.
2.2.2 If $(\tilde{k}_2, \tilde{k}_3) = (0, 0)$, we obtain a pair of **parallel planes** (**real** or **complex**), or a **double plane**, depending of the values of λ_1 and $\tilde{k}_1$.

Observe that in the imaginary cases, the complex numbers act as the ground field for the quadric. There, the designation complex or real makes no sense. For our purposes we deal with real surfaces, so we have decided to remark this difference.

As a conclusion, we can group together all the previous information in the following result.

Proposition 4.5.2 *Any quadric can be written, after an adequate orthogonal change of coordinates, in the form of one of the following implicit equations:*

- *Real and imaginary ellipsoid. Real ellipsoid: $\frac{x^2}{a^2} + \frac{y^2}{b^2} + \frac{z^2}{c^2} = 1$, for some $a, b, c \in \mathbb{R}^\star$. Imaginary ellipsoid: $\frac{x^2}{a^2} + \frac{y^2}{b^2} + \frac{z^2}{c^2} = -1$, for some $a, b, c \in \mathbb{R}^\star$.*

- *Hyperboloids of one and two sheets: Hyperboloid of one sheet: $\frac{x^2}{a^2} + \frac{y^2}{b^2} - \frac{z^2}{c^2} = 1$, for some $a, b, c \in \mathbb{R}^\star$. Hyperboloid of two sheets: $\frac{x^2}{a^2} + \frac{y^2}{b^2} - \frac{z^2}{c^2} = -1$, for some $a, b, c \in \mathbb{R}^\star$.*

- *Real and imaginary cone. Real cone: $\frac{x^2}{a^2} + \frac{y^2}{b^2} - \frac{z^2}{c^2} = 0$, for some $a, b, c \in \mathbb{R}^\star$. Imaginary cone: $\frac{x^2}{a^2} + \frac{y^2}{b^2} + \frac{z^2}{c^2} = 0$, for some $a, b, c \in \mathbb{R}^\star$.*

- *Elliptic and hyperbolic paraboloid: Elliptic paraboloid: $\frac{x^2}{a^2} + \frac{y^2}{b^2} + z = 0$, for some $a, b \in \mathbb{R}^\star$. Hyperbolic paraboloid: $\frac{x^2}{a^2} - \frac{y^2}{b^2} + z = 0$, for some $a, b \in \mathbb{R}^\star$.*

- *Real and imaginary elliptic cylinder: Elliptic cylinder: $\frac{x^2}{a^2} + \frac{y^2}{b^2} - 1 = 0$, for some $a, b \in \mathbb{R}^\star$. Imaginary elliptic cylinder: $\frac{x^2}{a^2} + \frac{y^2}{b^2} + 1 = 0$, for some $a, b \in \mathbb{R}^\star$.*

- *Hyperbolic cylinder: $\frac{x^2}{a^2} - \frac{y^2}{b^2} - 1 = 0$, for some $a, b \in \mathbb{R}^\star$.*

- *Parabolic cylinder: $x^2 + by + cz + d = 0$, for some $b, c \in \mathbb{R}$.*

- *Pairs of planes: Secant planes $\frac{x^2}{a^2} - \frac{y^2}{b^2} = 0$, for some $a, b \in \mathbb{R}^\star$; Parallel real planes $x^2 - a^2 = 0$, for some $a \in \mathbb{R}^\star$; Parallel imaginary planes $x^2 + a^2 = 0$, for some $a \in \mathbb{R}^\star$; Double plane $x^2 = 0$.*

With respect to the parametric representation of a quadric given in implicit form, it is always possible to achieve a parametrization of the quadric by means of the following algorithm.

Algorithm (Parametrization of a Quadric Surface)
Input: The implicit equation of a quadric, $F(x, y, z) = 0$.
Output: A parametrization of the quadric.

1. Compute the matrix representation of the quadric surface.
2. Classify the quadric, regarding its invariants.
3. Make appropriate changes of coordinates as described above (translations and isometries). The equation of the quadric in the new system of coordinates is determined in Proposition 4.5.2.
4. Distinguish the following cases to provide a parametrization of the quadric:

- Real ellipsoid $\frac{x^2}{a^2} + \frac{y^2}{b^2} + \frac{z^2}{c^2} = 1$ can be parametrized by

$$\begin{cases} x(t, s) = a\sin(t)\cos(s) \\ y(t, s) = b\sin(t)\sin(s) \\ z(t, s) = c\cos(t), \quad t \in \mathbb{R}, s \in \mathbb{R}. \end{cases}$$

  Observe that the whole surface except for a point is covered for $(t, s) \in (0, \pi) \times (0, 2\pi)$.

- Hyperboloid of one sheet $\frac{x^2}{a^2} + \frac{y^2}{b^2} - \frac{z^2}{c^2} = 1$ can be parametrized by

$$\begin{cases} x(t, s) = a\cosh(t)\cos(s) \\ y(t, s) = b\cosh(t)\sin(s) \\ z(t, s) = c\cosh(t), \quad t \in \mathbb{R}, s \in \mathbb{R}. \end{cases}$$

- Hyperboloid of two sheets $\frac{x^2}{a^2} + \frac{y^2}{b^2} - \frac{z^2}{c^2} = -1$ can be parametrized by

$$\begin{cases} x(t, s) = a\sinh(t)\cos(s) \\ y(t, s) = b\sinh(t)\sin(s) \\ z(t, s) = \pm c\cosh(t), \quad t \in \mathbb{R}, s \in \mathbb{R}. \end{cases}$$

  Each choice of the sign in the previous parametrization draws one or the other branches of the hyperboloid.

- The real cone $\frac{x^2}{a^2} + \frac{y^2}{b^2} - \frac{z^2}{c^2} = 0$, is parametrized by

$$\begin{cases} x(t, s) = as\cos(t) \\ y(t, s) = bs\sin(t) \\ z(t, s) = \pm cs, \quad t \in \mathbb{R}, s > 0. \end{cases}$$

  Each choice of the sign above draws the elements of the cone on one or the other side of the vertex in the cone. There is a half-line in the cone which is not parametrized when considering intervals of opening 2π on the domain of definition of the first parameter.

- The elliptic paraboloid $\frac{x^2}{a^2} + \frac{y^2}{b^2} + z = 0$, is parametrized by

$$\begin{cases} x(t, s) = as\cos(t) \\ y(t, s) = bs\sin(t) \\ z(t, s) = -s^2, \quad t \in \mathbb{R}, s > 0. \end{cases}$$

A curve contained in the elliptic paraboloid (half a parabola) is not parametrized when t is restricted to an open interval of length 2π.

- The hyperbolic paraboloid $\frac{x^2}{a^2} - \frac{y^2}{b^2} + z = 0$, is parametrized by

$$\begin{cases} x(t, s) = as\sinh(t) \\ y(t, s) = bs\cosh(t) \\ z(t, s) = s^2, \quad t \in \mathbb{R}, s \in \mathbb{R}. \end{cases}$$

and

$$\begin{cases} x(t, s) = as\cosh(t) \\ y(t, s) = bs\sinh(t) \\ z(t, s) = -s^2, \quad t \in \mathbb{R}, s \in \mathbb{R}. \end{cases}$$

Observe that the first of the previous parametrizations draws the hyperbolic paraboloid over the plane floor, whereas the second covers the points at negative height. The only point at the floor plane belonging to the paraboloid which is parametrized above is the origin.

- The elliptic cylinder $\frac{x^2}{a^2} + \frac{y^2}{b^2} - 1 = 0$, is parametrized by

$$\begin{cases} x(t, s) = a\cos(t) \\ y(t, s) = b\sin(t) \\ z(t, s) = s, \quad t, s \in \mathbb{R}. \end{cases}$$

A line contained in the cylinder is not parametrized when t is restricted to an open interval of length 2π.

- The hyperbolic cylinder $\frac{x^2}{a^2} - \frac{y^2}{b^2} - 1 = 0$, is parametrized by

$$\begin{cases} x(t, s) = a\cosh(t) \\ y(t, s) = b\sinh(t) \\ z(t, s) = s, \quad t, s \in \mathbb{R}. \end{cases}$$

- The parabolic cylinder $x^2 + by + cz + d = 0$, is parametrized by

$$\begin{cases} x(t, s) = t \\ y(t, s) = s \\ z(t, s) = (-d - t^2 - bs)c^{-1}, \quad t, s \in \mathbb{R} \end{cases}$$

in the case that $c \neq 0$; and

$$\begin{cases} x(t, s) = t \\ y(t, s) = (-d - t^2 - cs)b^{-1} \\ z(t, s) = s, \quad t, s \in \mathbb{R} \end{cases}$$

if $b \neq 0$.

- In case the quadric is a pair of planes or a double plane, each of them is of the form $\alpha x + \beta y + \gamma z = \delta$ for some $\alpha, \beta, \gamma, \delta \in \mathbb{R}$, and can be parametrized with one of the following:

$$\begin{cases} x(t, s) = t \\ y(t, s) = s \\ z(t, s) = (-\alpha t - \beta s + \delta)\gamma^{-1}, \quad t, s \in \mathbb{R} \end{cases} \qquad (\gamma \neq 0)$$

$$\begin{cases} x(t, s) = t \\ y(t, s) = (-\alpha t - \gamma s + \delta)\beta^{-1} \qquad (\beta \neq 0) \\ z(t, s) = s, \quad t, s \in \mathbb{R}. \end{cases}$$

$$\begin{cases} x(t, s) = (-\beta t - \gamma s + \delta)\alpha^{-1} \\ y(t, s) = t \qquad (\alpha \neq 0) \\ z(t, s) = s, \quad t, s \in \mathbb{R}. \end{cases}$$

5. Undo the change of coordinates.

The parametrizations above are based on the classic hyperbolic and trigonometric properties. The essence of spherical coordinates and cylindrical coordinates lies behind the parametrization of most of the quadrics in the previous classification.

Corollary 4.5.3 *Every ellipsoid and hyperboloid admits a parametrization whose components are given in terms of a linear combination of elements in*

$$\{1, \sin(t), \cos(t), \sinh(t), \cosh(t)\} \times \{1, \sin(s), \cos(s), \sinh(s), \cosh(s)\}.$$

If only trigonometric functions appear, then the surface is an ellipsoid.

The following example sheds light on the previous results.

Example 4.5.4 We consider the quadric of equation

$$-3 + 2x + 2y - 2xy + 2xz - y^2 - 2yz - z^2 = 0. \tag{4.12}$$

Its associated matrices are

$$M = \begin{pmatrix} -3 & 1 & 1 & 0 \\ 1 & 0 & -1 & 1 \\ 1 & -1 & -1 & -1 \\ 0 & 1 & -1 & -1 \end{pmatrix} \qquad M_0 = \begin{pmatrix} 0 & -1 & 1 \\ -1 & -1 & -1 \\ 1 & -1 & -1 \end{pmatrix}.$$

The eigenvalues of M_0 are $\lambda_1 = \sqrt{2}$, $\lambda_2 = -\sqrt{2}$ and $\lambda_3 = -2$. Also, $\det(M)/\det(M_0) = -7/4$, which means that the quadric is a hyperboloid of two sheets, in light of Case 1.1.4 above.

An associated orthonormal basis of eigenvectors is given by

$$\left\{ \left(\frac{\sqrt{2}}{2}, -\frac{1}{2}, \frac{1}{2} \right), \left(-\frac{\sqrt{2}}{2}, -\frac{1}{2}, \frac{1}{2} \right), \left(0, \frac{\sqrt{2}}{2}, \frac{\sqrt{2}}{2} \right) \right\}.$$

Proceeding with the orthogonal transformation gives rise to the change of coordinates

$$x = \frac{\sqrt{2}}{2} x_1 - \frac{\sqrt{2}}{2} y_1$$

$$y = -\frac{1}{2} x_1 - \frac{1}{2} y_1 + \frac{\sqrt{2}}{2} z_1$$

$$z = \frac{1}{2} x_1 + \frac{1}{2} y_1 + \frac{\sqrt{2}}{2} z_1$$

followed by the following translations, described in the corresponding part of the theory

$$x_1 = x_2 - \frac{\alpha}{\lambda_1}$$

$$y_1 = y_2 - \frac{\beta}{\lambda_2}$$

$$z_1 = z_2 - \frac{\gamma}{\lambda_3}$$

for $\alpha = \frac{1}{2}(\sqrt{2} - 1)$, $\beta = \frac{1}{2}(-\sqrt{2} - 1)$ and $\gamma = \frac{\sqrt{2}}{2}$. In the last coordinate system, the equation of the quadric is

$$\sqrt{2}x_2^2 - \sqrt{2}y_2^2 - 2z_2^2 = \frac{7}{4},$$

which can be parametrized by

$$x_2(t, s) = \pm a \cosh(t)$$
$$y_2(t, s) = b \sinh(t) \cos(s)$$
$$z_2(t, s) = c \sinh(t) \sin(s), \qquad t \in \mathbb{R}, s \in \mathbb{R}.$$

Note that the whole hyperboloid except for a curve in it is parametrized for every interval of length 2π with respect to the parameter s. The inverse changes of coordinates yields the following parametrizations of the hyperboloid:

$$x(t, s) = -\frac{\sqrt{14}}{4} \sinh(t) \cos(s) \pm \frac{\sqrt{7}}{4} \cosh(t) + \frac{1}{2}$$

$$y(t, s) = \frac{\sqrt{7}}{4} \sinh(t) \sin(s) - \frac{\sqrt{7}}{4} \sinh(t) \cos(s) + \frac{3}{4} \mp \frac{\sqrt{14}}{8} \cosh(t)$$

$$z(t, s) = \frac{\sqrt{7}}{4} \sinh(t) \sin(s) + \frac{\sqrt{7}}{4} \sinh(t) \cos(s) - \frac{1}{4} \pm \frac{\sqrt{14}}{8} \cosh(t) \qquad (4.13)$$

Figure 4.39 illustrates the quadric considered in the example.

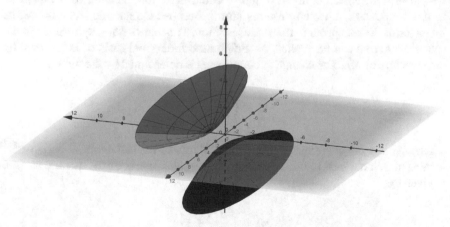

Fig. 4.39 Quadric defined in Eq. (4.12) and parametrized by Eq. (4.13)

Some elements of the quadric surfaces can be directly found by inspection of their representing matrix. In what follows we describe some examples of such elements.

Definition 4.5.5 A point in $\mathbb{R}^3$ is a **center** of a quadric if it is a center of symmetry for the quadric. The **axis** of quadric with center are the symmetry axes of that quadric.

Proposition 4.5.6 *Let $C = (c_1, c_2, c_3)$ be a center of the quadric surface represented by the matrix*

$$M = \left(\begin{array}{c|c} d & b^T \\ \hline b & M_0 \end{array} \right).$$

Then it holds that the quadric admits a center if and only if $rank(M_0) = rank(M_0|b)$. In addition to that, every center of the quadric is determined by the solution of the system

$$M_0 \begin{pmatrix} x \\ y \\ z \end{pmatrix} = -b.$$

The argument in the proof of Proposition 3.1.9 can be adapted here.

In light of Proposition 4.5.6, and the previous classification of quadrics, the quadrics with a center are: the ellipsoid, hyperboloid of one or two sheets, the cone, elliptic and hyperbolic cylinders, and secant, parallel or double planes.

Let us deduce the equations of the axes of a quadric with a unique center. First, we perform a translation of the origin of coordinates to the center of the quadric. In this framework, the vector b turns into 0. Because the quadric has a center, its linear terms do not appear in this representation. The matrix M_0 is symmetric so the spectral theorem can be applied, obtaining an orthonormal basis of $\mathbb{R}^3$ formed by eigenvectors of M_0. The change of coordinates is determined by the matrix

$$Q = \left(\begin{array}{c|c} 1 & O \\ \hline 0 & Q_0 \end{array} \right),$$

where Q_0 has the coordinates of the eigenvectors of the orthonormal basis of $\mathbb{R}^3$ located at its columns. In this coordinate system, the matrix of the quadric surface is given by

$$M' = \left(\begin{array}{c|c} d & O \\ \hline 0 & M_0' \end{array} \right),$$

with $M_0' = \mathrm{diag}(\lambda_1, \lambda_2, \lambda_3)$, and the quadric is given by

$$\lambda_1 (x^{**})^2 + \lambda_2 (y^{**})^2 + \lambda_3 (z^{**})^2 + d = 0$$

in the new coordinate system. The coordinate axes $\{x^{**} = y^{**} = 0\}$, $\{x^{**} = z^{**} = 0\}$ and $\{y^{**} = z^{**} = 0\}$ determine the axes of the quadric. We undo the change of coordinates performed to arrive at the equations of the axes of the quadric with center:

$$(x, y, z) = (c_1, c_2, c_3) + t u_1,$$

$$(x, y, z) = (c_1, c_2, c_3) + t u_2,$$

$$(x, y, z) = (c_1, c_2, c_3) + t u_3,$$

where $C = (c_1, c_2, c_3)$ is the center of the quadric, and $\{u_1, u_2, u_3\}$ is the orthonormal basis of eigenvectors of $\mathbb{R}^3$ with respect to Q_0.

Some of the quadrics are ruled surfaces. All reducible quadrics are ruled surfaces; the hyperbolic paraboloid is a double ruled surface, and so the hyperboloid of one sheet.

In Example 4.4.3, we have shown the construction of a hyperboloid of one sheet by means of two families of lines.

4.6 Quadrics Revisited: Some Examples in Architecture

Now let us apply the study of quadric surfaces in Sect. 4.5 to specific manifestations of these surfaces in architectural elements. We will mainly focus on ellipsoids, hyperboloids and paraboloids. Examples of such surfaces in architecture could fill a whole book, so we have only selected a few examples.

At first sight, Sheraton Huzhou Hot Spring Resort, designed by Ma Yansong (see Fig. 4.40) might look like a toroidal structure. However, it seems more likely to model an ellipsoid from which an elliptic cylinder has been removed.

Other structures inspired in this surface are the ellipsoid for the United Nations Geospatial Information Management Forum in Deqing County, China, designed by the Architectural Design & Research Institute of Zhejiang University, directed by the architects Zhenling Wu and Bing Chen, located at Huzhou City; the Infosys Building in Hinjewadi, Pune, India, by Hafeez Contractor and M/S Construction; and Taiwan Centers for Disease Control Buildings, by Studio Nicoletti Associati. We also refer to the National Center for the Performing Arts in Beijing (see Fig. 3.21).

In the particular case of an sphere, numerous structures have appealed to this surface for lighting, structural or aesthetic reasons. It is well-known that the inside of the Pantheon in Rome resembles a sphere (Sperling 2015). More recently, spherical

Fig. 4.40 Sheraton hotel

Fig. 4.41 Apple

shapes have inspire sculptures and other urban installments throughout the world. Examples of these are Avicii Arena, by Svante Berg and Lars Vretbla, located at Stockholm, Sweden; Apple Marina Bay Sands in Singapore by Foster + Partners (Fig. 4.41), the Academy Museum of Motion Pictures, in Los Angeles, USA, by

Fig. 4.42 James S. McDonnell Planetarium by Gyo Obata

Renzo Piano, and many others. We refer to the webpage (Dezeen 2021b) for more examples.

Concerning hyperboloid structures, the one-sheet hyperboloid is a recurrent surface in architecture due to its double ruled nature, which provides strength and stability to the building. It can be found in towers such as the water tower in Ciechanow (already mentioned for its toroidal tank); the water tower in Les Essarts-le-Roi in France; Canton tower in Guangzhou, and many others. It also characterizes the well-known cooling tower of nuclear reactors.

In this direction, one can cite the Cathedral of Brasilia by Oscar Niemeyer (see Fig. 1.23); James S. McDonnell Planetarium by Gyo Obata located in Forest Park in St. Louis (see Fig. 4.42); and remarkable buildings by Eduardo Torroja such as the water tank in Fedala or the Zarzuela's Hippodrome by Carlos Arniches, Martín Domínguez, and Eduardo Torroja. The buildings by Eduardo Torroja have been analysed in detail in Torroja (1962b,a), and other articles in that book are dedicated to his oeuvre. A more general insight is shown in Birindelli and Cedrone (2012).

The application of the elliptic paraboloid in satellite communications is due to the geometric properties of a parabola, which concentrates in a single point all the incident rays in perpendicular position. Hyperbolic paraboloids, as ruled surfaces, provide suitable structural answers to shapes in buildings, leading to both parabolas and hyperbolas from families of lines, and a special curvature. Examples of them in architecture are L'Oceanogràfic designed by by Félix Candela in Valencia; Los Manantiales Restaurant at Xochimilco, Mexico, also by Felix Candela, and other works of the same architect such as his well-known umbrellas and roof shells; the public library of Tromsø, designed by Kjell Beiteby; Scandinavium by

Fig. 4.43 Scotiabank Saddledome by GEC Architecture

Poul Hultberg, in Gotheburg, Sweden, and the Scotiabank Saddledome by GEC Architecture, in Calgary (Fig. 4.43).

A list of hyperboloid structures can be found in Wikipedia (2019b). A nice example of this structure applied in architecture was Laboratorios Jorba (nicknamed The Pagoda), by Miguel Fisac (Barrallo and Sánchez-Beitia 2011) which contains a nice review on organic architecture by Miguel Fisac as well as Torroja and Candela.

4.7   Curvature: Minimal and Developable Surfaces

The concept of curvature at a point on a surface and its consequences and constructions are enormously important and rich. Here, we only treat them briefly in order to provide some practical examples. For a deeper study in this direction, we refer to Costa et al. (1997), do Carmo (1976).

In Chap. 1 (see (1.12) and the subsequent constructions) and Chap. 2 (see (2.9) and the subsequent constructions) we considered the notion of curvature of a regular plane and spatial curve. This notion can be generalized to the framework of surfaces. Let us consider a surface parametrized by (U, X). It is natural to think about the curvature of a surface at a point in terms of the curvature of the regular curves contained in the surface and passing through that point. We recall this was the case when defining the tangent plane of a regular surface at a point (see Definition 3.1.16). Let us consider a regular curve (I, α) contained in the surface and such that (I, α) is a natural parametrization of the curve. We write $\alpha(t) = X(u(t), v(t))$, for $t \in I$. In view of (2.9) we obtain the curvature $\kappa_\alpha(t)$.

Definition 4.7.1 The projection of $\alpha''(t)$ on the normal vector (3.6) associated to the surface at the point $\alpha(t)$, i.e.,

$$n(t) := \frac{\frac{\partial X}{\partial u}(u(t), v(t)) \times \frac{\partial X}{\partial v}(u(t), v(t))}{\left\| \frac{\partial X}{\partial u}(u(t), v(t)) \times \frac{\partial X}{\partial v}(u(t), v(t)) \right\|},$$

is called the **normal curvature vector** at $\alpha(t)$ with respect to (I, α), and notated by $\tilde{k}_n(t)$. The **tangent curvature** vector is given by $\alpha''(t) - k_n(t)$. The **normal curvature** of the surface (U, X) at $\alpha(t)$ is defined by

$$k_n(t) = \alpha''(t) \cdot n(t),$$

i.e. $\tilde{k}_n(t) = k_n(t)n(t)$.

Analogous definitions can be obtained for curves not defined by a natural parametrization.

The **second fundamental form** associated to a regular surface parametrized by (U, X) is represented by the matrix

$$\begin{pmatrix} \frac{\partial^2 X}{\partial u^2}(u, v) \cdot n(u, v) & \frac{\partial^2 X}{\partial u \partial v}(u, v) \cdot n(u, v) \\ \frac{\partial^2 X}{\partial u \partial v}(u, v) \cdot n(u, v) & \frac{\partial^2 X}{\partial v^2}(u, v) \cdot n(u, v) \end{pmatrix},$$

where $n(u, v)$ denotes the normal vector associated to the surface S at the point $X(u, v)$ defined in Eq. (3.6). Observe that the elements in the previous matrix correspond to the projections of the second order derivatives of X onto the line at the point under study and direction given by the normal vector. The eigenvalues of the previous matrix are real due to the spectral theorem. We usually refer to the second fundamental form to the quadratic form defined on the vectors of $T_S(P)$ (see Definition 3.1.16) whose representation in the basis of $\mathbb{R}^2$ given by

$$\left\{ \frac{\partial X}{\partial u}(u, v), \frac{\partial X}{\partial v}(u, v) \right\} \tag{4.14}$$

is the previous matrix. We observe that given a vector $\omega = (\omega_1, \omega_2, \omega_3)$ in $T_S(P)$ of coordinates (x_1, x_2) in the basis of Eq. (4.14), i.e.,

$$(\omega_1, \omega_2, \omega_3) = x_1 \frac{\partial X}{\partial u}(u, v) + x_2 \frac{\partial X}{\partial v}(u, v),$$

then the second fundamental form of v is given by

$$II(\omega) = \begin{pmatrix} x_1 & x_2 \end{pmatrix} \begin{pmatrix} \frac{\partial^2 X}{\partial u^2}(u, v) \cdot n(u, v) & \frac{\partial^2 X}{\partial u \partial v}(u, v) \cdot n(u, v) \\ \frac{\partial^2 X}{\partial u \partial v}(u, v) \cdot n(u, v) & \frac{\partial^2 X}{\partial v^2}(u, v) \cdot n(u, v) \end{pmatrix} \begin{pmatrix} x_1 \\ x_2 \end{pmatrix}. \tag{4.15}$$

Proposition 4.7.2 *Let (U, X) be a regular surface, and let (I, α) be a regular parametrization of a curve contained in it. It holds that*

$$k_n(\alpha'(t)) = \frac{II(\alpha'(t))}{I(\alpha'(t), \alpha'(t))}. \tag{4.16}$$

The previous definition does not depend on the curve (I, α) with a common tangent line considered, due to Meusnier's theorem (see Costa et al. (1997), do Carmo (1976)), so the normal curvature at a point of a regular surface can be associated to any direction in the tangent plane. The sign of $II(v)$ coincides with that of the normal curvature, which determines whether it is definite or not, seen as a quadratic form. In this regard we have that

- If $\det(II) > 0$ in a vicinity of the point under study, then II is positive or negative definite and it turns out that the surface lies on one side of the tangent plane near the point. In this case, we say the point is an **elliptic point**.
- If $\det(II) < 0$ in all vicinities of the point under study, then II is not definite and there exist points in the surface on both sides of the tangent plane no matter which neighborhood of the point one considers. In this case, we say the point is a **hyperbolic point**.
- If $\det(II) = 0$, then except for the points in one direction, the surface lies on one side of the tangent plane near the point. In this case, we say the point is a **parabolic point**.

The so-called **principal directions** of S at a point $P \in S$ are those vectors in $T_S(P)$ for which the function k_n attains its maximum and minimum. It is straightforward to verify that a vector $\omega = (\omega_1, \omega_2, \omega_3) \in T_S(P)$ with coordinates (x_1, x_2) in the basis of $T_S(P)$ given by Eq. (4.14) is a principal direction if both

$$\frac{\partial k_n(\omega)}{\partial x_1} = 0, \qquad \frac{\partial k_n(\omega)}{\partial x_2} = 0$$

hold. Regarding Eq. (4.16), the previous equalities are satisfied if and only if

$$\begin{vmatrix} (x_1)^2 & -x_1 x_2 & (x_1)^2 \\ \frac{\partial^2 X}{\partial u^2}(u, v) \cdot n(u, v) & \frac{\partial^2 X}{\partial u \partial v}(u, v) \cdot n(u, v) & \frac{\partial^2 X}{\partial v^2}(u, v) \cdot n(u, v) \\ E & F & G \end{vmatrix} = 0.$$

A practical way to obtain the principal directions at a point is to consider the second order equation in λ:

$$\xi_1 \lambda^2 + \xi_2 \lambda + \xi_3 = 0, \tag{4.17}$$

with

$$\xi_1 = G \frac{\partial^2 X}{\partial u \partial v}(u, v) \cdot n(u, v) - F \frac{\partial^2 X}{\partial v^2}(u, v) \cdot n(u, v),$$

$$\xi_2 = G \frac{\partial^2 X}{\partial u^2}(u, v) \cdot n(u, v) - E \frac{\partial^2 X}{\partial v^2}(u, v) \cdot n(u, v),$$

$$\xi_3 = F \frac{\partial^2 X}{\partial u^2}(u, v) \cdot n(u, v) - E \frac{\partial^2 X}{\partial u \partial v}(u, v) \cdot n(u, v).$$

Let λ_1 and λ_2 be the two solutions of (4.17). Then the principal directions are given by $(1, \lambda_1)$ and $(1, \lambda_2)$. The curvatures in such directions, i.e.,

$$k_1 = k_n((1, \lambda_1)), \quad \text{and} k_2 = k_n((1, \lambda_2))$$

are known as the **principal curvatures** at P.

We conclude this part by defining Gaussian and mean curvatures.

Definition 4.7.3 Under the same assumptions and notations above, we define Gaussian curvature at a point of the surface by $K = k_1 k_2$. We define **mean curvature** at a point of the surface by $H = (k_1 + k_2)/2$.

Recent studies indicate that a constant mean curvature is desired in architectural structures, see Tellier et al. (2018).

At this point, we define minimal and developable surfaces as follows.

Definition 4.7.4 A **minimal surface** is a surface with zero mean curvature at all its points.

Example 4.7.5 Let us consider a catenoid, consisting of the surface of revolution (see Sect. 4.4) determined by the rotation of a catenary (see Sect. 1.2) around the axis $\{x = y = 0\}$. We consider the parametrization of the catenary in the XZ plane given by

$$\alpha(t) = (a \cosh(t/a), 0, t), \quad t \in \mathbb{R},$$

for some $a > 0$ (see Eq. (1.6)). Therefore, the surface of revolution is determined by the parametrization

$$X(u, v) = \left(a \cosh(\frac{v}{a} \cos(u)), a \cosh(\frac{v}{a}) \sin(u), v \right), \quad u \in (0, 2\pi), v \in \mathbb{R}.$$

For every $X(u, v)$ in the catenoid, one has

$$\frac{\partial X}{\partial u}(u, v) = (-a \cosh(v/c) \sin(u), a \cosh(v/c) \cos(u), 0),$$

$$\frac{\partial X}{\partial v}(u, v) = (\sinh(v/a) \cos(u), \sinh(v/a) \sin(u), 1),$$

and

$$n(u, v) = \left(\frac{\cos(u)}{\cosh(v/a)}, \frac{\sin(u)}{\cosh(v/a)}, -\frac{\sinh(v/a)}{\cosh(v/a)} \right).$$

The terms in the first and second fundamental form are computed from the previous elements which yield

$$E = a^2 \cosh^2(v/a), \quad F = 0, \quad G = \cosh^2(v/a),$$

and

$$II(\omega) = (x_1 \; x_2) \begin{pmatrix} -a & 0 \\ 0 & 1/a \end{pmatrix} \begin{pmatrix} x_1 \\ x_2 \end{pmatrix}.$$

We conclude from the previous expression and Eq. (4.16) that

$$k_n(\omega_1, \omega_2) = \frac{-a^2\omega_1^2 + \omega_2^2}{a \cosh^2(v/a)(a^2\omega_1^2 + \omega_2^2)}.$$

This function of two variables attains its minimum and maximum values at the vectors $(1, 0)$ and $(0, 1)$, and therefore

$$k_1 = -\frac{1}{a \cosh^2(v/a)}, \quad k_2 = \frac{1}{a \cosh^2(v/a)}.$$

This means that $H = 0$, and the catenoid is a minimal surface. Observe that we have also obtained the directions of the vectors of minimum and maximum curvature at any point of the surface.

Figure 4.44 illustrates a catenoid for $a = 1$. It is worth mentioning the modelling of vaulted structures studied in Nikolić and Živaljević (2020).

Example 4.7.6 Let us consider a surface of revolution (see Sect. 4.4) determined by the rotation of a plane curve lying in the XZ plane rotating around the vertical axis $\{x = y = 0\}$. We assume that the plane curve is parametrized as a graph of a regular function (see Proposition 3.1.9)

$$\alpha(u) = (f(u), 0, u), \quad u \in I,$$

for some open interval $I \subseteq \mathbb{R}$ and $f(u) > 0$ for all $u \in I$. According to the results in Sect. 4.4, the surface of revolution can be parametrized by

$$X(u, v) = (f(u)\cos(v), f(u)\sin(v), u), \quad u \in \mathbb{R}, v \in I.$$

We aim to search for minimal surfaces of this nature.

The computation of the first and second fundamental forms yield Eq. (4.16), with

$$k_n((\omega_1, \omega_2)) = -\frac{9f(u)(\omega_1^2 f''(u) - \omega_2^2 f(u))}{|f(u)|\sqrt{1 + f(u)^2}(\omega_1^2 + \omega_2^2)(u^2 + v^2 + 1)^2},$$

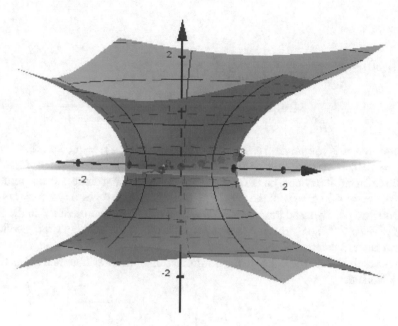

Fig. 4.44 Catenoid of Example 4.7.5, with $a = 1$

for every vector (ω_1, ω_2) in the tangent plane at $X(u, v)$. We have

$$\frac{\partial k_n}{\partial \omega_1}(\omega_1, \omega_2) = -\frac{18(f''(u) + f(u))\omega_2^2 f(u)\omega_1}{|f(u)|\sqrt{1 + f(u)^2}(\omega_1^2 + \omega_2^2)(u^2 + v^2 + 1)^2},$$

and

$$\frac{\partial k_n}{\partial \omega_2}(\omega_1, \omega_2) = \frac{18(f''(u) + f(u))\omega_1^2 f(u)\omega_2}{|f(u)|\sqrt{1 + f(u)^2}(\omega_1^2 + \omega_2^2)(u^2 + v^2 + 1)^2}.$$

They are both zero if $f''(u) + f(u) = 0$ for every $(u, v) \in I \times \mathbb{R}$. In this case, one does not arrive at a minimal surface after the substitution of f'' in $k_n(\omega_1, \omega_2)$. If $(u, v) \in I \times \mathbb{R}$ is such that $f(u) = 0$, then we have that the points $X(u, v)$ belong to the axis of rotation. This irregular situation has been dismissed in the hypotheses because the point is not a regular point of the surface. We also consider the vectors (ω_1, ω_2) such that

- $\omega_1 = 0$.
- $\omega_2 = 0$.

We arrive at

$$k((1,0)) = -\frac{9f(u)f''(u)}{|f(u)|\sqrt{1+(f'(u))^2}(u^2+v^2+1)^2},$$

$$k((0,1)) = \frac{9f(u)f(u)}{|f(u)|\sqrt{1+(f'(u))^2}(u^2+v^2+1)^2}. \qquad (4.18)$$

We find that it is a minimal surface if $k((1,0)) + k((0,1)) = 0$, i.e., if $f''(u) + f(u) = 0$. This second order differential equation has $f(u) = C_1 e^u + C_2 e^{-u}$ as solution, for arbitrary constants $C_1, C_2 \in \mathbb{R}$. In order that there are no points of intersection of the curve with the axis $\{x = y = 0\}$ then $C_1 C_2 > 0$. We may assume that both are positive, and moreover, after a translation, one can write f in the form of $ce^{u/c} + ce^{-u/c}$, i.e., $f(u) = c \cosh(u/c)$. Therefore, the surface is the revolution of a catenary: a catenoid.

This result can also be linked to the minimization of the area obtained in Eq. (3.3), which leads to

$$\left\| \frac{\partial X}{\partial u}(u,v) \times \frac{\partial X}{\partial v}(u,v) \right\| = f(u)\sqrt{1+f'(u)^2}.$$

Therefore, Eq. (3.3) reads as follows

$$2\pi \int_{u_1}^{u_2} f(u)\sqrt{1+f'(u)^2} du,$$

which is minimum among the regular functions such that $f(u_1) = x_1$ and $f(u_2) = x_2$ with $f(u)$ being a catenary arc (see Sagan (1992)).

In this direction, we refer to the workshop and study described in Mackin (2016).

Definition 4.7.7 A **developable surface** is a surface with zero Gaussian curvature at all its points.

Example 4.7.8 Any plane is a minimal and developable surface. Let us consider a plane of equation $ax + by + cz = d$, for some $a, b, c, d \in \mathbb{R}$. Without loss of generality, we assume that $a \neq 0$ (otherwise, we follow the same procedure with respect to the variable whose coefficient does not vanish). A parametrization of the plane is

$$X(u,v) = \left(d - \frac{b}{a}u - \frac{c}{a}v, u, v \right), \quad u, v \in \mathbb{R}.$$

It turns out that the normal vector $n(u, v)$ associated to the previous parametrization at $X(u, v)$ is defined in terms of

$$\frac{\partial X}{\partial u}(u, v) = (-\frac{b}{a}, 1, 0), \quad \frac{\partial X}{\partial v}(u, v) = (-\frac{c}{a}, 0, 1).$$

Therefore,

$$n(u, v) = \frac{1}{\sqrt{1 + (b^2 + c^2)/a^2}} \left(1, \frac{b}{a}, \frac{c}{a}\right),$$

and $II(\omega) \equiv 0$ for all $\omega \in T_S(X(u, v))$. This means that $k_n \equiv 0$, so $k_1 = k_2 = 0$, and the Gaussian and mean curvatures are null.

Observe from the previous definitions that the plane is the only surface which is both developable and minimal.

At this point, we are ready to study these two types of surfaces from the point of view of architecture. Generally speaking, a minimal surface is a surface which minimizes its area locally. They were first linked to the surface of minimum surface which satisfied some boundary conditions (an illustrative image could be the form of a bubble in a soap bubble toy with the shape of a closed spatial curve). The advances in geometric design make it possible to obtain complex structures based on or inspired by these surfaces. Concerning architectural constructions, one may refer to Emmer (2013) and also Velimirovic et al. (2008), where these kind of surfaces are treated from the mathematical point of view and display some applications in architecture.

One of the most outstanding architects who studied and was inspired by minimal surfaces in his work is Frei Otto. His tensile and membrane structures rely on the action of forces leading to surfaces closely related to families of curves associated to a catenary. Also, some of his experiments included the study of shape of soap bubbles, closely related to minimal surfaces. Figure 4.45 shows Frei Otto's Olympiastadion in Munich as a realization of this type of architecture. We also refer to his design for German pavilion for the International and Universal Exposition, held in Montreal in 1967, among many others. We refer to Glaeser (1972) for a broader and in-depth study of his buildings. We also cite Emmer (2015) for an architectural view of soap bubbles and soap films.

Other buildings inspired on this phenomena are Beijing National Aquatics Center, by PTW Architects, CSCEC, CCDI, and Arup, which follows Plateau's rules in soap form. Several ideas for buildings have been inspired by the Enneper surface, which is a minimal surface studied in Example 4.7.9. As an example, in the International Conference FORM and FORCE- Structural Membranes- 2019, held in Barcelona, a work on minimal surface tensegrity networks in the case of an Enneper surface pavilion structure was presented. The abstract is available in Liapi et al. (2019). Minimal surfaces have also been of interest to sculptors (Séquin 2008).

Fig. 4.45 Olympiastadion in Munich by Frei Otto

Example 4.7.9 An Enneper surface is parametrized by

$$X(u, v) = (\frac{u}{3}(1 - \frac{u^2}{3} + v^2), -\frac{v}{3}(1 - \frac{v^2}{3} + u^2), \frac{u^2 - v^2}{3}),$$

for $u, v \in \mathbb{R}$. It is not difficult to verify that

$$II(\omega) = A(u, v) \begin{pmatrix} x_1 \\ x_2 \end{pmatrix},$$

with

$$A(u, v) = \begin{pmatrix} x_1 & x_2 \end{pmatrix} \begin{pmatrix} p_1(u, v) & 0 \\ 0 & p_1(u, v) \end{pmatrix},$$

with $p_1(u, v) = \frac{1}{9}(1 + 2u^2 + u^4 + 2v^2 + 2u^2v^2 + v^4)$ and

$$I(\omega, \omega) = \begin{pmatrix} x_1 & x_2 \end{pmatrix} \begin{pmatrix} -\frac{2}{3} & 0 \\ 0 & \frac{2}{3} \end{pmatrix} \begin{pmatrix} x_1 \\ x_2 \end{pmatrix}.$$

for all tangent vector ω to the surface at the point $X(u, v)$ with coordinates (x_1, x_2) in the coordinate system determined by the partial derivatives of the parametrization, as in the previous examples.

These results can be incorporated into Eq. (4.16) to get that

$$k_n(\omega_1, \omega_2) = -\frac{6(\omega_1^2 - \omega_2^2)}{(u^4 + 2u^2v^2 + v^4 + 2u^2 + 2v^2 + 1)(\omega_1^2 + \omega_2^2)}.$$

Writing the previous function in polar coordinates, one may write

$$k_n(\rho, \theta) := C(u, v)\cos(2\theta), \quad \text{with } C(u, v) = \frac{-6}{u^4 + 2u^2v^2 + v^4 + 2u^2 + 2v^2 + 1}.$$

Therefore, this function attains its maximum and minimum values at $\theta = -\pi/4$ and $\theta = \pi/4$ respectively, and consequently

$$k_1 = -\frac{6}{u^4 + 2u^2v^2 + v^4 + 2u^2 + 2v^2 + 1},$$

$$k_2 = \frac{6}{u^4 + 2u^2v^2 + v^4 + 2u^2 + 2v^2 + 1}$$

are the extreme values of the curvature at $X(u, v)$. We have that $H = 0$ so an Enneper surface is a minimal surface. Note that the directions of minimum and maximum curvature have also been obtained.

Figure 4.46 displays this surface.

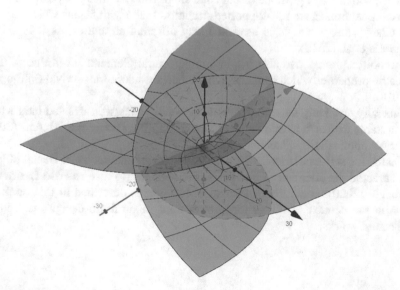

Fig. 4.46 An Enneper surface

Now, we give some brief information about developable surfaces and their application in architecture. The importance of these surfaces is that they can be flattened onto a plane maintaining lengths in the representation. From the opposite point of view, an incompressible sheet can produce such surface by bending. Moreover, all of them belong to the larger class of ruled surfaces; a hyperboloid of one sheet is a ruled surface which is not developable.

The oloid and the sphericon are (see Sect. 4.1), cylinders and the cone excluding from the vertex are developable surfaces.

In a more practical framework, recent advances have been made in developable surfaces and related topics in architectural studies. Several works included in AAG (2008) illustrate this. We only give some examples included in this book. In Kilian et al. (2008) (see also the references therein) the authors develop a computational way to reconstruct and design surfaces which can be obtained by curved folding. The authors of Tang et al. (2016) deal with the analysis and design of developable surfaces in freeform architecture. We also refer to Koschitz et al. (2008), where the authors study a subfamily of developable surfaces from the point of view of origami, as surfaces coming from a sheet of paper; and also Schiftner et al. (2008) or from a different approach in Postle (2012). Both developable and minimal surface are also studied in Schling et al. (2018).

4.7.1   Final Comments

This brief section aims to comment on other mathematical aspects in architectural elements, which are of great interest but that are beyond the main scope of the book.

Reciprocal frames are self-supported structures made with beams which support each other. This technique is used in many different structures, as it is shown in Barrallo et al. (2018).

Voronoi diagrams have also inspired architectural elements, even due to their geometric properties which animals such as cheetahs adopt for survival (Vergopoulos 2010).

Generally speaking, tensegrity structures usually consist in isolated bars which remain stable thanks to tensile forces (Motro 2003). We also refer to Marcus (2008) for further applications in architecture.

Finally, the self-similarity properties and interesting geometry of fractal objects have attracted the attention of architects (Sala 2003). We also refer to Lastra and de Miguel (2020) where the mathematical techniques described in this book are shown in the context of the current trends used in furniture design inspired by architectural works.

4.8 Suggested Exercises

4.1. Find the parametrization of the cylindrical surface of directrix d and generatrices which are parallel to the line r, where

$$d \equiv \begin{cases} y^3 = x^5 \\ z = 0 \end{cases} \qquad r \equiv \begin{cases} x + 2y = 0 \\ x + y + z = 0. \end{cases}$$

4.2. Find a parametrization of the conical surface with vertex at the point $P = (1, 1, -1)$ and directrix given by the line of equations

$$\begin{cases} z = 3 \\ \frac{x^2}{9} + \frac{y^2}{4} = 1. \end{cases}$$

Find its implicit equation. Is it a regular surface?

4.3. Find the tangent developable surface associated to the curve parametrized by $(\mathbb{R}, \alpha)$, with

$$\alpha(t) = (\exp(t), \exp(-t), \exp(t^2)), \quad t \in \mathbb{R}.$$

4.4. Let S be the quadric

$$S = \{(x, y, z) \in \mathbb{R}^3 : yz - x^2 = 0\}.$$

Verify it is a cone and that $(\mathbb{R}^2, X)$ is a parametrization of this surface, with

$$X(u, v) = (u(v + 1), u^2(v + 1), v + 1), \quad u, v \in \mathbb{R}.$$

In addition to this, verify that such parametrization corresponds to the conical
surface with vertex at $P = (0, 0, 0)$ and directrix given by $C = \{(x, y, z) \in \mathbb{R}^3 : y - x^2 = 0\}$.

4.5. Consider the surface parametrized by

$$X(u, v) = (u, v, (1 - \frac{v}{4}) \sin(\frac{u}{3}) + \frac{v}{7}), \quad (u, v) \in \mathbb{R}^2.$$

Obtain an implicit representation of this surface, and the tangent plane at the
point $P = (3\pi, 7, 1)$. The surface above describes a ruled surface. Verify that
the segment parametrized by

$$(x, y, z) = (3\pi, 7t, t), \quad t \in \mathbb{R}$$

is contained in the surface.

4.6. Consider the quadric of equation

$$1 + 2x - 2y + x^2 + y^2 + z^2 + 4yz = 0.$$

Fig. 4.47 QR Code 20

Classify the quadric and, if the following makes sense, find its center. In case it is unique, find the change of coordinates which make the quadric has the simplest form.

4.7. Consider the quadric of equation

$$-1 - 4y + x^2 + 2y^2 + 3z^2 = 0.$$

Classify the quadric and, if it makes sense to do so, find its center. In case it is unique, find the change of coordinates which make the quadric have the simplest form.

4.8. Let $m \in \mathbb{R}$. Consider the quadric of equation

$$x^2 + y^2 + z^2 + 2mxy - 1 = 0.$$

Classify the quadric depending on the different values of $m \in \mathbb{R}$. The QR Code of Fig. 4.47 links to the continuous deformation of the quadric with respect to the parameter.

4.9. Determine the parametrization of an oloid from its geometric representation. Do the same for the sphericon.

4.10. Check that the helicoid

$$X(u, v) = (u \cos(v), u \sin(v), v), \quad (u, v) \in I \times \mathbb{R},$$

is a minimal surface. Here, $I = (\alpha, \beta)$ stands for an open interval, with $0 < \alpha < \beta$.

4.11. Obtain the parametric description of the surface generated by the revolution of an ellipse, i.e., substitute the moving circle in the construction of a toroidal surface by an ellipse.

4.12. Solve the same problem by substituting the circle at the floor plane by an ellipse.

4.13. Let S be the conoidal surface generated by joining the points of the circle

$$C = \{(x, y, z) \in \mathbb{R}^3 : x^2 + z^2 = 1, y = h, z \geq 0\},$$

for some $h > 0$, with the segment $L = \{(t, 0, 0) : t \in [0, 1]\}$. The rule to join points is that the points in C and L share their first component. Give both parametric and implicit descriptions of the surface.

4.14. Sweeping: Construct the pseudosphere as the revolution around the OZ axis of the curve parametrized by

$$t \mapsto \left(\frac{2}{e^t + e^{-t}}, 0, t - \frac{e^{2t} - 1}{e^{2t} + 1} \right)$$

(a tractrix) for $t \in \mathbb{R}$. Check that its Gaussian curvature is a negative constant for all points of the pseudosphere.

4.15. Let (I, α) be a regular surface. For every $t \in I$, we consider the regular plane curve (I_t, β_t), which depends on the element $t \in I$. Moreover, for every $t \in I$, we fix a point $\beta_t(s_t) \in \beta_t(I_t)$.

We consider the parametrization of the surface constructed as follows: for every $t \in I$, let $(I_t, \tilde{\beta}_t)$ be the regular plane curve obtained by the translation of $\beta_t(s_t)$ to $\alpha(t)$, and a rigid movement which transforms the plane in which the plane curve $\beta_t(I_t)$ is contained into the plane with normal vector $T_\alpha(t)$ (preserving the choice of the orientation for every $t \in I$).

Figure 4.48 illustrates the situation in the case where $(I = \mathbb{R}, \alpha)$ is the circle $\alpha(t) = (\cos(t), \sin(t), 1)$ and for every $t \in \mathbb{R}$, $(I_t = \mathbb{R}, \beta_t)$ consists of the parabola $\beta_t(s) = -s^2$, with $\beta(s_t)$ being the vertex of the parabola.

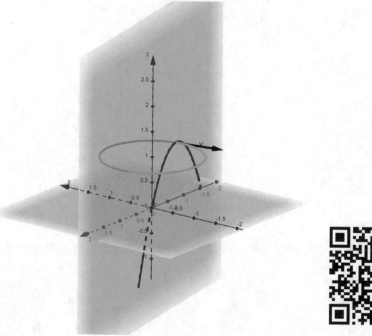

Fig. 4.48 Example of curve shifting. QR Code 21

The QR Code of Fig. 4.48 links to the construction of the surface.

Give another example of the construction of a surface following the previous technique, departing from a curve (I, α), with $\alpha_3 \equiv C$, for some $C \in \mathbb{R}$.

Appendix A
Coordinate Systems

This appendix presents a brief review of different coordinate systems in $\mathbb{R}^2$ and $\mathbb{R}^3$.

Regarding the two-dimensional affine space we only outline two of the most important coordinate systems. A coordinate system in $\mathbb{R}^2$ consists of an origin of coordinates, which is a distinguished point in the plane, and two elements which allow us to represent every other point in the plane uniquely in terms of those elements.

Cartesian Coordinates in $\mathbb{R}^2$

The two elements mentioned above are two orthonormal vectors in this case. Any point in $\mathbb{R}^2$ is represented by their Cartesian coordinates, which are the coefficients of the linear combination of the vector in the fixed orthonormal basis. The choice of the canonical basis in $\mathbb{R}^2$, $\{(1, 0), (0, 1)\}$, causes that the coordinates of a vector to coincide with the vector itself:

$$(x_0, y_0) = (0, 0) + x_0(1, 0) + y_0(0, 1).$$

The set of vectors $\{\left(\frac{\sqrt{2}}{2}, \frac{\sqrt{2}}{2}\right), \left(\frac{\sqrt{2}}{2}, -\frac{\sqrt{2}}{2}\right)\}$ and the point $P = (x_1, y_1) \in \mathbb{R}^2$ determine another Cartesian coordinate system in $\mathbb{R}^2$ (Fig. A.1).

The point $(x_0 y_0)$ has coordinates determined by

$$\left(\frac{x_0 + y_0 - x_1 - y_1}{\sqrt{2}}, \frac{x_0 - y_0 - x_1 + y_1}{\sqrt{2}}\right),$$

because

$$(x_0, y_0) = (x_1, y_1) + \frac{x_0 + y_0 - x_1 - y_1}{\sqrt{2}}\left(\frac{\sqrt{2}}{2}, \frac{\sqrt{2}}{2}\right) + \frac{x_0 - y_0 - x_1 + y_1}{\sqrt{2}}\left(\frac{\sqrt{2}}{2}, -\frac{\sqrt{2}}{2}\right).$$

© The Author(s), under exclusive license to Springer Nature Switzerland AG 2021
A. Lastra, *Parametric Geometry of Curves and Surfaces*, Mathematics and the Built
Environment 5, https://doi.org/10.1007/978-3-030-81317-8

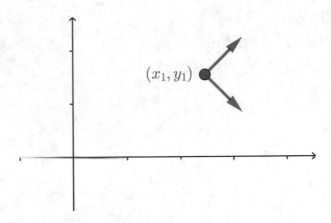

Fig. A.1 A Cartesian coordinate system

Regarding Definition 1.3.11, the set $\{\alpha(t), \{T_\alpha(t), N_\alpha(t)\}\}$ is a reference system of the same nature as the previous ones, for every $t \in I$, with (I, α) as a regular parametrization.

Polar Coordinate System

A point $(x_0, y_0) \neq (0, 0)$ in Cartesian coordinates can be uniquely written in polar coordinates (Fig A.1). This system consists in considering both the distance to the origin of the point and the angle with respect to the OX axis in the range $[0, 2\pi)$. Therefore, the point (x_0, y_0) has polar coordinates (ρ, θ), with $\rho = \sqrt{x_0^2 + y_0^2} > 0$ and θ is the angle just mentioned. The inverse change of coordinates is

$$(x_0, y_0) = (\rho \cos(\theta), \rho \sin(\theta)).$$

This coordinate system is quite useful when studying spiral-like curves or other curves whose parametrization might depend on the angle and/or the distance to the origin (Fig. A.2).

$$t \mapsto (\rho(t) \cos(\theta(t)), \rho(t) \sin(\theta(t))),$$

for some functions $\rho, \theta : I \to \mathbb{R}, \rho(t) > 0$.

Example A.0.1 The implicit equation of a logarithmic spiral in polar coordinates is determined as follows. We consider the parametrization

$$\alpha(t) = (a \exp(bt) \cos(t), a \exp(bt) \sin(t)),$$

which can be rewritten in terms of ρ and θ, with

$$\rho := \sqrt{a^2 \exp(2bt) \cos^2(t) + a^2 \exp(2bt) \sin^2(t)} = |a| \exp(bt), \qquad \theta := t,$$

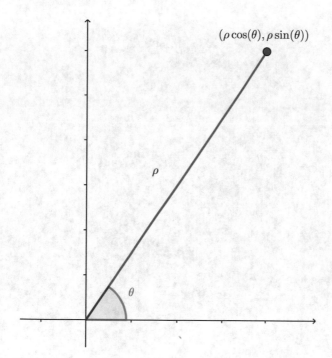

$(\rho\cos(\theta), \rho\sin(\theta))$

ρ

θ

Fig. A.2 Polar coordinate system

which means that the implicit definition of the curve is

$$\{(\rho, \theta) \in (0, \infty) \times \mathbb{R} : \rho - |a|e^{b\theta} = 0\}.$$

The elements (curvature, arc length, etc) related to a parametrization have their own formulation in this coordinate system.

Another coordinate system is the log-polar coordinate system, analogous to the previous one, but measuring $\log(\rho)$, instead of ρ, so that the restriction of being positive disappears.

Example A.0.2 The implicit equation of a logarithmic spiral in the log-polar coordinate system is determined as follows. We consider $\log(\rho)$ instead of ρ in the implicit equation of the previous example, i.e.,

$$\{(\tilde{\rho}, \theta) \in \mathbb{R} \times \mathbb{R} : \tilde{\rho} - \log(|a|) - b\theta = 0\}.$$

Cartesian Coordinates in $\mathbb{R}^3$

We describe the context of points belonging to Euclidean space. There exists a one-to-one mapping associating each point with its coordinates.

Cartesian coordinates are the most frequently used coordinate system in Euclidean space. These coordinates describe the distance from a point to each of the

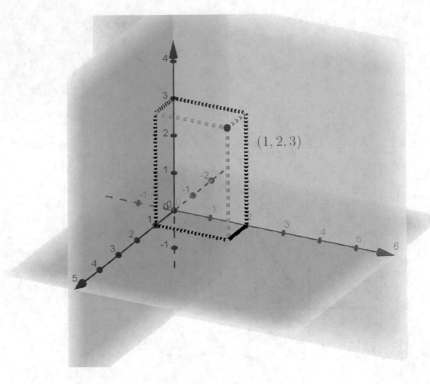

Fig. A.3 Cartesian coordinates of the point $(1, 2, 3)$

coordinate planes. In Fig. A.3, we illustrate the representation of the point $(1, 2, 3)$ in Cartesian coordinates. Any point (x_0, y_0, z_0) in Euclidean space is written in the form

$$(x_0, y_0, z_0) = (0, 0, 0) + x_0(1, 0, 0) + y_0(0, 1, 0) + z_0(0, 0, 1).$$

This coordinate system is useful in architecture when referring to lines in space (see Fig. A.4). Any line is parametrized in these coordinates by

$$\alpha(t) = (x_0, y_0, z_0) + t(v_1, v_2, v_3), \quad t \in \mathbb{R},$$

where $P = (x_0, y_0, z_0)$ is a point of the line, and $v = (v_1, v_2, v_3)$ is a directing vector of the line.

Cylindrical coordinate system has implicitly appeared in Sect. 2.5. This coordinate system combines the polar coordinate system with the Cartesian coordinate system. The third component of this system is preserved, whereas given a fixed height, the point is determined by polar coordinates inside that plane.

Fig. A.4 Lines in the Gran
Via Capital Hotel, Spain by
La Hoz Arquitectura

The point (x_0, y_0, z_0) has cylindrical coordinates given by (ρ, θ, z_0), where $\rho = \sqrt{x_0^2 + y_0^2}$ and $\theta \in [0, 2\pi)$ is the angle $\theta = \arctan(\frac{y_0}{x_0})$.

The coordinates representing a point in $\mathbb{R}^3$ in Cartesian coordinates determine uniquely the cylindrical coordinates of the same point (Fig. A.5):

$$\begin{cases} x = \rho \cos(\theta) \\ y = \rho \sin(\theta) \\ z = \quad z. \end{cases}$$

We have already observed in the previous section that this type of coordinate system is adequate when considering space curves of spiralling nature, such as helices.

A third widespread coordinate system is the **spherical coordinate system**. This system makes use of polar coordinates twice. Given a point in Euclidean space, one first determines in polar coordinates its projection on the XY plane. The angle obtained is notated by φ. We consider $0 \leq \varphi < 2\pi$. Next, one considers the plane determined by the origin of coordinates, the point, and the projection (if the point lies inside the XY plane, we consider that plane) and measure the angle departing

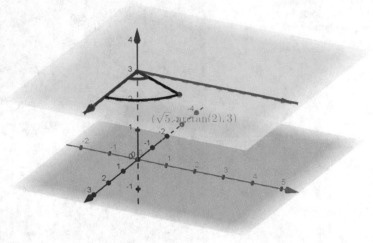

Fig. A.5 Cylindrical coordinates of the point $(1, 2, 3)$

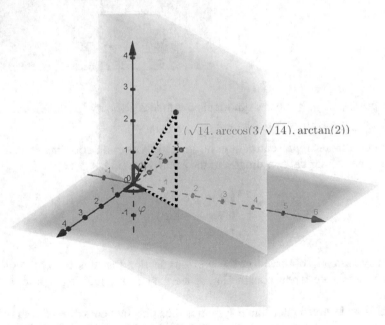

Fig. A.6 Spherical coordinates of the point $(1, 2, 3)$

from the positive points of the OZ axis. That angle is denoted by θ; we have $0 \leq \theta < \pi$. Finally, $r \geq 0$ stands for the distance of the point to the origin of coordinates. The spherical coordinates of that point are represented by (r, θ, φ). Figure A.6 illustrates this consideration for the point $(1, 2, 3)$.

The coordinates which represent a point in $\mathbb{R}^3$ in Cartesian coordinates can be written in terms of spherical coordinates in the following way

$$\begin{cases} x = r\sin(\theta)\cos(\varphi) \\ y = r\sin(\theta)\sin(\varphi) \\ z = \quad r\cos(\theta). \end{cases}$$

The reciprocal relation is given by

$$\begin{cases} r = \quad \sqrt{x^2 + y^2 + z^2} \\ \theta = \arccos\left(\frac{z}{\sqrt{x^2+y^2+z^2}}\right) \\ \varphi = \quad \arctan\left(\frac{y}{x}\right). \end{cases}$$

Such coordinates are useful when dealing with sphere-like surfaces such as the Cloud Gate in Chicago by Anish Kapoor (Fig. A.7). We also mention the geodesic domes such as those by Buckminster Fuller, or some deployable structures by Emilio Pérez Piñero.

Fig. A.7   Cloud Gate in Chicago by Anish Kapoor

The changes of coordinates from one to other system can be performed by taking into account the nature of their elements. Numerous other coordinate systems in Euclidean space can be named, including parabolic cylindrical, paraboloidal, elliptic cylindrical or toroidal coordinates, among many others.

Appendix B
Mathematical Tool Kit

This appendix provides an overview of the definitions and results appearing in this book. They can be found in many books on linear algebra, calculus in one and several variables, and differential equations.

Our aim is to briefly present or review the main notions, rather than be exhaustive, so the results are just mentioned from an intuitive point of view and the proofs of the results are omitted.

B.1 Introduction to Linear Algebra

B.1.1 Systems of Linear Equations

A system of linear equations is a finite set of linear equations of the form

$$\begin{cases} a_{11}x_1 + a_{12}x_2 + \ldots + a_{1n}x_n = b_1 \\ a_{21}x_1 + a_{22}x_2 + \ldots + a_{2n}x_n = b_2 \\ \vdots \\ a_{m1}x_1 + a_{m2}x_2 + \ldots + a_{mn}x_n = b_m, \end{cases}$$

for certain real numbers a_{ij}, known as coefficients, and b_j, known as independent terms. If all $b_{ij} = 0$, the system is said to be a homogeneous system. A solution of the previous system is a set of numbers $\alpha_1, \ldots, \alpha_n$ which satisfies all the m equations. Observe that an homogeneous system always admits $x_1 = \cdots = x_n = 0$ as a solution. A system might not have any solution, or might have one or an infinite

© The Author(s), under exclusive license to Springer Nature Switzerland AG 2021
A. Lastra, *Parametric Geometry of Curves and Surfaces*, Mathematics and the Built Environment 5, https://doi.org/10.1007/978-3-030-81317-8

number of solutions. A system of linear equations can be represented in matrix form
by

$$
\begin{pmatrix}
a_{11} & a_{12} & \cdots & a_{1n} \\
a_{21} & a_{22} & \cdots & a_{2n} \\
\vdots & \vdots & \ddots & \vdots \\
a_{m1} & a_{m2} & \cdots & a_{mn}
\end{pmatrix}
\begin{pmatrix}
x_1 \\
x_2 \\
\vdots \\
x_n
\end{pmatrix}
=
\begin{pmatrix}
b_1 \\
b_2 \\
\vdots \\
b_m
\end{pmatrix}.
$$

Gaussian elimination is a procedure which allows us to transform any given system
of linear equations into another system in which the matrix of coefficients is an
upper triangular matrix, i.e., with null elements under the main diagonal, preserving
the set of solutions of the system. The Rouché-Frobenius theorem determines the
number of solutions of a system in terms of the rank of the matrix of coefficients A
and the extended matrix $(A|b)$, which adds the independent terms as its last column.

B.1.2   Vector Spaces

A vector space is a non-empty set on which are defined two operations, whose
elements, known as vectors, satisfy certain properties. In the present book, we only
consider real vector spaces, mainly finitely generated. This means that this vector
space is essentially $\mathbb{R}^n$ for some $n \geq 1$. A subspace of $\mathbb{R}^n$ is a subset $\emptyset \neq W \subseteq \mathbb{R}^n$
which satisfies the condition that any linear combination of vectors in W remains
in W. In fact, a subspace is a set which maintains the structure of the initial vector
space in the set of vectors forming the subset.

A set of n vectors in $\mathbb{R}^n$ which are linearly independent can generate the whole
space $\mathbb{R}^n$. This means that any other vector in $\mathbb{R}^n$ can be written as a linear
combination of them, and this linear combination is unique. Given a subspace
$W \subseteq \mathbb{R}^n$, one can also consider sets of vectors which are linearly independent
and generate the whole subspace by means of linear combination of their elements.

The number of elements in any of such sets remains constant, and it is known as
the dimension of the subspace. As a matter of fact, $\mathbb{R}^n$ is a vector space of dimension
n. The canonical basis of $\mathbb{R}^n$ consists of the set of vectors

$$
\{(1, 0, 0, \ldots, 0), (0, 1, 0, \ldots, 0), \ldots, (0, \ldots, 1)\}.
$$

Given a subspace $A \subseteq \mathbb{R}^n$, we can identify all of its elements provided that a
basis in $\mathbb{R}^n$ is fixed. This fixed basis usually is chosen to coincide with the canonical
basis for practical reasons. Indeed, a subspace W of $\mathbb{R}^n$ can be determined as
the vectors whose components satisfy an homogeneous system of linear equations
(implicit form of W). Also, the elements of W can be determined by the evaluation
at certain number of parameters (parametric form of W).

We can consider linear maps between two vector spaces, or homomorphisms. This kind of mapping preserves linear combinations of vectors. Again, in our framework we deal with linear mappings of the form $f : \mathbb{R}^n \rightarrow \mathbb{R}^m$, for some $m, n \geq 1$. This definition means that $f(\lambda u + \mu v) = \lambda f(u) + \mu f(v)$ for every $\lambda, \mu \in \mathbb{R}$ and all $u, v \in \mathbb{R}^n$.

A homomorphism can be represented by a matrix once a basis has been fixed in the vector spaces involved. In our framework, we initially fix the canonical basis in $\mathbb{R}^n$ and $\mathbb{R}^m$. Given a linear mapping $f : R^n \rightarrow \mathbb{R}^m$, the set of all vectors in $\mathbb{R}^n$ which are sent to the null vector forms a subspace of $\mathbb{R}^n$, known as the kernel of f. The set of all vectors of $\mathbb{R}^m$ which are the image of some vector in $\mathbb{R}^n$ is also a subspace of $\mathbb{R}^m$, known as the image of f.

B.1.3   Euclidean Vector Spaces

A Euclidean vector space is a vector space endowed with a scalar product. A scalar product in $\mathbb{R}^n$ is a map $\langle \cdot, \cdot \rangle : \mathbb{R}^n \rightarrow \mathbb{R}$ which satisfies the following properties for all $u, v, w \in \mathbb{R}^n$ and $\lambda \in \mathbb{R}$:

- $\langle u, v \rangle = \langle v, u \rangle$
- $\langle u + v, w \rangle = \langle u, w \rangle + \langle v, w \rangle$
- $\langle \lambda \cdot u, v \rangle = \lambda \langle u, v \rangle$
- $\langle u, u \rangle \geq 0$.

As an example, the canonical scalar product in $\mathbb{R}^n$ is defined by

$$\langle (x_1, \ldots, x_n), (y_1, \ldots, y_n) \rangle := x_1 y_1 + \ldots + x_n y_n.$$

The concept of angle between two vectors emerges from this notion. As a matter of fact, the Gram-Schmidt process is a method which determines an orthonormal basis for any scalar product (in this case of $\mathbb{R}^n$). This means that we can obtain a basis $\{u_1, \ldots, u_n\}$ of $\mathbb{R}^n$ such that $\langle u_i, u_j \rangle = 0$ for $1 \leq i, j \leq n$ with $i \neq j$ and $\langle u_i, u_i \rangle = 1$ for $1 \leq i \leq n$. In the present work, this process is mainly used to "move" any conic (resp. quadric) in $\mathbb{R}^2$ (resp. $\mathbb{R}^3$) without causing deformations to the conic (resp. quadric). A scalar product in $\mathbb{R}^n$ is usually represented by a $n \times n$ matrix, after fixing a basis $\{u_1, \ldots, u_n\}$ in $\mathbb{R}^n$. The coefficient at the position (i, j) in that matrix is determined by $\langle u_i, i_j \rangle$.

B.1.4   Diagonalization: Eigenvalues and Eigenvectors

Given an $n \times n$ matrix with real coefficients, or equivalently, a linear mapping from the vector space $\mathbb{R}^n$ to $\mathbb{R}^n$ with the canonical basis (or any other basis) fixed, we can search for another basis in which the matrix representation of the linear map is in diagonal form. This is not always possible to find but it is if the matrix is symmetric, due to the spectral theorem. The vectors which form a basis in which a diagonal representation of a linear map is determined are known as eigenvectors of A, and are collected as the linearly independent vectors $v = (v_1, \ldots, v_n) \in \mathbb{R}^n$ which satisfy the condition that

$$A - \lambda v^T = (0, \ldots, 0)^T,$$

for some $\lambda \in \mathbb{R}$, known as the eigenvalue associated to the eigenvector v.

B.2   Real Functions of One Variable

Many global features can be studied when dealing with a function $f : A \to \mathbb{R}$, for $\emptyset \neq A \subseteq \mathbb{R}$, such as its domain (values where the function is well defined), sections by the coordinate axes, symmetry, or asymptotes (those lines to which the graph of a function gets close when approaching infinity), etc.

The local study of a function $f : (a, b) \to \mathbb{R}$ at any $c \in (a, b) \subseteq \mathbb{R}$ (with possibly $a = -\infty$ or $b = +\infty$) leans on the concept of continuity and differentiability. Generally speaking, a function is continuous at $c \in (a, b)$ if the limits of the images of values close to c approach $f(c)$, as much as needed. A function which is continuous at $c \in (a, b)$ is said to be differentiable at c if

$$\lim_{x \to c} \frac{f(x) - f(c)}{x - c}.$$

exists and is a finite number, denoted by $f'(c)$. In the case where this is valid for all $c \in (a, b)$, we can talk about the derivative of f, which is a function $f' : (a, b) \to \mathbb{R}$, sending each value in (a, b) to the derivative of f at that point. This concept can be extended to f' arriving at f'', and so on. If a function admits n derivatives in (a, b), all of which are continuous, then we say that the function belongs to the space $C^n((a, b))$. If the function admits any number of derivatives in (a, b), the function is said to be C^∞ in (a, b).

The concept of derivative allows us to define the tangent line of a function at a point by

$$y - f(c) = f'(c)(x - c).$$

This line is the one which best fits the curve determined by the graph of f near the point c. The concept of the tangent line can be extended to a polynomial of some degree to the Taylor polynomial of a function at a point, if the adequate number of derivatives is well defined near the point under study (Fig. B.1). This polynomial determines a better approximation of the function near the point under study, and it

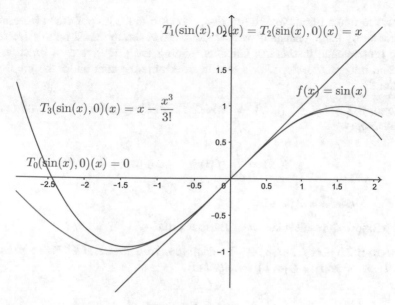

Fig. B.1 $f(x) = \sin(x)$ and some Taylor polynomials at $x = 0$

is defined by

$$T_n(f, c)(x) = f(c) + f'(c)(x - c) + \frac{f''(c)}{2!}(x - c)^2 + \ldots + \frac{f(n)(c)}{n!}(x - c)^n.$$

The derivative is the main tool in order to study the local behavior of a function near a point. The sign of the derivative at a point is the clue to determine if it is increasing or decreasing. It is also used to determine concavity and convexity of a function near the point.

Theorem B.2.1 *Let $f : (a, b) \to \mathbb{R}$ which admits $n - 1$ derivatives in (a, b), for some $n \geq 2$ and such that $f^{(n)}(c)$ exists. If*

$$f'(c) = f''(c) = \ldots = f^{(n-1)}(c) = 0$$

and $f^{(n)}(c) \neq 0$, then

- *If n is even, then f has a maximum near c if $f^{(n)}(c) > 0$ and a minimum near c if $f^{(n)}(c) < 0$.*
- *If n is odd, then f is strictly monotone increasing at c if $f^{(n)}(c) > 0$, and strictly monotone decreasing if $f^{(n)}(C) < 0$.*

Given a function $f : A \to \mathbb{R}$, the reciprocal concept of derivative of f is known as the primitive. Under certain conditions, there exists a function $F : A \to \mathbb{R}$ such that $F'(x) = f(x)$ for all $x \in A$. F is a primitive of f. There are several tools and methods to find a primitive of a function, such as the method of direct integration, integration by parts, change of variable, etc. However, the most impressive result is the fundamental theorem of calculus relating the primitive of a function and Riemann integral, developed as a tool to determine the area under the graph of a function.

Theorem B.2.2 *Let* $f : [a, b] \rightarrow \mathbb{R}$ *which admits primitive in* $[a, b]$, *and define the function*

$$F(x) = \int_a^x f(t)dt, \quad x \in [a, b].$$

The, F is continuous in $[a, b]$.

The following result is known as Barrow's rule.

Theorem B.2.3 *Let* $f : [a, b] \rightarrow \mathbb{R}$ *a continuous function, and let F be a primitive of f. Then for every* $x \in [a, b]$, *it holds that*

$$\int_a^x f(t)dt = F(x) - F(a).$$

The link between primitives and Riemann integral allows us to transfer the methods applied to primitives such as integration by parts or the change of variable to Riemann integrals. A Riemann integral is useful in many applications:

- to compute the area between the graphs of two functions (Fig. B.2). Given $f, g :$ $[a, b] \rightarrow \mathbb{R}$ which are integrable functions in $[a, b]$, the area between f and g from $x = a$ to $x = b$ is given by

$$\int_a^b |f(x) - g(x)|dx.$$

- to compute the volume/area of surfaces, such as surfaces of revolution. The volume of a surface of revolution generated by the function $y = f(x)$, with $f : [a, b] \rightarrow \mathbb{R}$ which turns around the OX axis is given by

$$\pi \int_a^b f^2(x)dx.$$

The area of that surface of revolution is given by

$$2\pi \int_a^b f(x)\sqrt{1 + (f'(x))^2}dx.$$

Fig. B.2 Area between
$f(x) = 2xe^{x^2} - 4x$ and OX
from $x = 0$ and $x = 1$

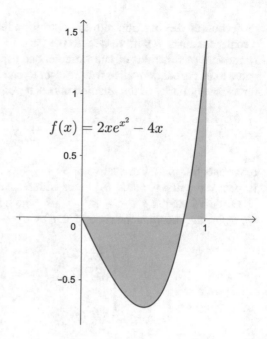

$$f(x) = 2xe^{x^2} - 4x$$

- to compute the arclength of the graph of a function. Given an integrable function
  $f : [a, b] \to \mathbb{R}$, the arclength of the curve between $x = a$ and $x = b$ is given by

$$\int_a^b \sqrt{1 + f'(x)^2} dx.$$

These applications are described in more detail in Sect. 1.3.

B.3   Functions of Several Real Variables

The treatment of functions of several real variables is different from that of functions
of one variable. The topology described by open intervals in the real line is
substituted by that of open discs in $\mathbb{R}^2$, or open spheres in $\mathbb{R}^3$. More generally,
an open set in $\mathbb{R}^n$ is a set such that the distance from each of its points to the
complementary set is positive. Given a function $f : A \subseteq \mathbb{R}^n \to \mathbb{R}^m$ for some
$n, m \geq 1$ the concept of continuity of f at a point $c \in A$ coincides with that of one
variable intuitively, concerning each of the components of $f = (f_1, \ldots, f_m)$, with
$f_j : A \to \mathbb{R}$ being a scalar function. However, the different approaches to get close
to c make it harder to study. Differentiation distinguishes derivatives with respect to
each variable. Indeed, the differentiation of f at c is equivalent to the differentiation
of each of the components at c. From the geometric point of view, a partial derivative
with respect to one variable of a scalar function can be seen as the derivative of the

projection of the function into the corresponding plane. The partial derivative of a scalar function f with respect to the variable x is denoted by $\partial f / \partial x$. As in the framework of functions of one variable, one can construct a function sending each value to its partial derivative with respect to one variable, and repeat this process as far as the regularity of the function makes it possible to proceed. For example,

$$\frac{\partial^m f}{\partial x_1^{m_1} \cdots \partial x_n^{m_n}}$$

represents the partial derivative $m = m_1 + \ldots + m_n$ times of f, m_1 times with respect to the first variable, m_2 times with respect to the second, etc.

Given a function $f : A \subseteq \mathbb{R}^n \to \mathbb{R}^m$ which admits partial derivatives at $c \in A$, we define the Jacobian matrix of f at c by

$$\begin{pmatrix} \frac{\partial f_1}{\partial x_1}(c) & \frac{\partial f_1}{\partial x_2}(c) & \cdots & \frac{\partial f_1}{\partial x_n}(c) \\ \frac{\partial f_2}{\partial x_1}(c) & \frac{\partial f_2}{\partial x_2}(c) & \cdots & \frac{\partial f_2}{\partial x_n}(c) \\ \vdots & \vdots & \ddots & \vdots \\ \frac{\partial f_m}{\partial x_1}(c) & \frac{\partial f_m}{\partial x_2}(c) & \cdots & \frac{\partial f_m}{\partial x_n}(c) \end{pmatrix}.$$

This is usually denoted by $\mathcal{J}_f(c)$. If $m = 1$ (i.e., f is a scalar function), this matrix is known as the gradient of f at c:

$$\nabla(f)(c) = \left(\frac{\partial f}{\partial x_1}(c), \frac{\partial f}{\partial x_2}(c), \ldots, \frac{\partial f}{\partial x_n}(c) \right).$$

Two equivalent results are of great importance in the study of functions of several variables, used in some of the theoretical statements in the present book. These are the implicit function theorem and the inverse function theorem.

Theorem B.3.1 (Implicit Function Theorem) *Let $\emptyset \neq A \subseteq \mathbb{R}^{n+m}$ be an open set, $f : A \to \mathbb{R}^m$ in $C^k(A)$ for some $k \geq 1$. Let $c = (a, b) \in A$, with $a \in \mathbb{R}^n$, $b \in \mathbb{R}^m$ such that $f(a, b) = 0$. We assume that the determinant*

$$det \left(\frac{\partial f_i}{\partial x_{n+j}}(a, b) \right)_{1 \leq i, j \leq m} \neq 0.$$

Then there exists $a \in U \subseteq \mathbb{R}^n$, $b \in V \subseteq \mathbb{R}^m$ open sets such that for all $x \in U$, there exists a unique $\varphi(x) \in V$ with $f(x, \varphi(x)) = 0$. Moreover, the function $\varphi : U \to V$ belongs to $C^k(U)$.

Generally speaking, the above result describes a condition that allows us to write some of the variables in terms of the other variables, at least locally.

Theorem B.3.2 (Inverse Mapping Theorem) *Let $\emptyset \neq A \subseteq \mathbb{R}^n$ be an open set, and let $f : A \to \mathbb{R}^n$ belong to $C^k(A)$ for some $k \geq 1$. We assume $det(\mathcal{J}_f(c)) \neq 0$,* *for some $c \in A$. Then there exist open sets $c \in V \subseteq A$, and $f(c) \in W \subseteq \mathbb{R}^n$ such that $f : V \to W$ is a one to one function and $f^{-1} : W \to V$ belongs to $C^k(W)$.*

In this book, we are working with functions which admit any number of partial derivatives at every point of the domain. This framework makes things easier in the sense that the concept of differentiability coincides with that of the existence of partial derivatives. In addition to this, the order in which the partial derivatives are computed does not change the final result (Schwarz's theorem). The role of tangent line in three dimensions is played by the tangent plane. The tangent plane of a scalar differentiable function $f : A \subseteq \mathbb{R}^3 \to \mathbb{R}$ at $(x, y, z) = c = (c_1, c_2, c_3)$ is

$$z - f(c) = f(c) + \frac{\partial f}{\partial x}(c)(x - c_1) + \frac{\partial f}{\partial y}(c)(y - c_2).$$

Also, the approximation via a Taylor polynomial can also be extended to the framework of several variables.

A procedure to find local maximum and minimum points of a scalar function of several real variables is the several-variable version of that of functions in one variable. Under the assumption that the function is sufficiently regular, we can search for the points with associated tangent plane parallel to the floor plane, i.e., the gradient evaluated at such points is zero. These are known as critical points. In order to classify each point, one makes use of the Hessian matrix. Assume that $c \in \mathbb{R}^n$ is a critical point of f. The Hessian matrix of $f : A \subseteq \mathbb{R}^n \to \mathbb{R}$ at $c \in \mathbb{R}^n$ is given by

$$\mathcal{H}f(c) = \begin{pmatrix} \frac{\partial^2 f}{\partial x_1^2}(c) & \cdots & \frac{\partial^2 f}{\partial x_1 \partial x_n}(c) \\ \vdots & \ddots & \vdots \\ \frac{\partial^2 f}{\partial x_n \partial x_1}(c) & \cdots & \frac{\partial^2 f}{\partial x_n^2}(c) \end{pmatrix}.$$

Sylvester's criterion determines a way to proceed at this point. Let $c \in A$ be a critical point of f. Let A_k be the determinant of the submatrix of $\mathcal{H}f(c)$ given by the elements in the first k rows and first k columns. If the signs of such determinants are all positive then f has a relative minimum at c. If the signs of such determinants vary with the sequence $(-1)^k$ then f has a relative maximum at c. Otherwise, the method does not give any information, and we should use another way to proceed.

B.4   Differential Equations and Systems of Differential Equations

An ordinary differential equation is an equation of the form $F(x, y, y', \cdots, y^{(n)}) = 0$, where F is a function of $n + 2$ variables, for some $n \in \mathbb{N}$. A solution to the ordinary differential equation is a function $y = y(x)$ which admits n derivatives and satisfies the equation in certain domain. There are many kinds of differential equations classified by the maximum number of derivatives of the unknown function y involved (the order) and other issues. A differential equation of the form

$$\frac{dy}{dx} + a(x)y = b(x)$$

for some functions a, b is known as a linear differential equation of first order. In this work, a system of three linear differential equations appear in the framework of existence of space curves having a prescribed curvature and torsion. A general form of such a system is given by

$$\begin{cases} y_1' = a_{11}(x)y_1 + a_{12}(x)y_2 + \ldots + a_{1n}(x)y_n + b_1(x) \\ \vdots \\ y_n' = a_{n1}(x)y_1 + a_{n2}(x)y_2 + \ldots + a_{nn}(x)y_n + b_n(x) \end{cases}$$

where $a_{i,j}(x), b_j(x)$ are continuous functions defined on some common interval I of the real line. Any solution of this problem is a vector space of dimension n. The Picard-Lindelöf theorem guarantees that this system has a unique solution provided the prescribed values of $y_1(c), y_2(c), \ldots, y_n(c)$, for some $c \in I$.

Appendix C
Solution to the Suggested Exercises

C.1 Chapter 1

1.1. Any point $P = (x, y) \in \mathbb{R}^2$ which belongs to the set of points is such that

$$\text{dist}(P, P_1)^2 + \text{dist}(P, P_2)^2 = 6,$$

or equivalently

$$(x + \sqrt{2})^2 + y^2 + (x - \sqrt{2})^2 + y^2 = 6.$$

The simplification of the previous expression yields $x^2 + y^2 = 1$. Therefore, the set of points coincides with the circle centered at the origin, and unit radius.

The substitution of the property in the statement of the problem yields the line of equation $x = 3\sqrt{2}/4$.

1.2. It is straightforward to verify that the derivatives of every order of the function f can be obtained at every point $(x, y) \in \mathbb{R}^2$. In addition to this, one has

$$\frac{\partial^{j+k}}{\partial x^j \partial y^k} f(x, y) = (-1)^k \exp(x - y).$$

1.3. It is straightforward to verify that the functions $p_1 \equiv 1$, $p_2 \equiv x$ and $p_3 \equiv y$ belong to $\mathcal{C}^\infty(\mathbb{R}^2)$. Any other polynomial p can be written in the form

$$p(x, y) = a_{(m,n)}x^m y^n + a_{(m-1,n)}x^{m-1} y^n + a_{(m,n-1)}x^m y^{n-1} + \ldots + a_{(0,0)},$$

for some $a_{(j,k)} \in \mathbb{R}$, and some non-negative integers m, n. Then p a regular function because it is the composition of regular functions in $\mathbb{R}^2$.

© The Author(s), under exclusive license to Springer Nature Switzerland AG 2021
A. Lastra, *Parametric Geometry of Curves and Surfaces*, Mathematics and the Built
Environment 5, https://doi.org/10.1007/978-3-030-81317-8

1.4. Let $f : I \times \mathbb{R} \to \mathbb{R}$ be defined by $f(x, y) = y - g(x)$. For every $(x_0, y_0) \in I \times \mathbb{R}$ one has

$$\nabla f(x_0, y_0) = \left(\frac{\partial f}{\partial x}(x_0, y_0), \frac{\partial f}{\partial y}(x_0, y_0) \right) = (-g'(x_0), 1) \neq (0, 0).$$

Theorem 1.1.7 can be applied to guarantee that the graph of g is a regular curve. The parametrization (I, α) defined by $\alpha(t) = (t, g(t))$ parametrizes the whole curve.

1.5. It holds that α is a one-to-one function, defining a regular arc near each of the points in its range. Moreover, $\alpha'(t) = (2t, 3t^2) = (0, 0)$ if and only if $t = 0$. This corresponds to the point $(0, 0)$ of the curve known as a cusp displayed in Fig. C.1.

1.6. We recall that the unique singular point of the lemniscate is the origin of coordinates. Regarding its implicit definition as the zeroes of the function $f(x, y) = (x^2 + y^2)^2 - xy$, we find that the tangent line at $P = (x_0, y_0) \neq (0, 0)$ has equation

$$(4x_0(x_0^2 + y_0^2) - y_0)(x - x_0) + (4y_0(x_0^2 + y_0^2) - x_0)(y - y_0) = 0.$$

Fig. C.1 Cusp. Exercise 1.5

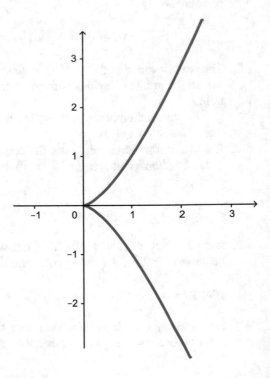

The normal line is given by

$$(x_0 - 4y_0(x_0^2 + y_0^2))(x - x_0) + (4x_0(x_0^2 + y_0^2) - y_0)(y - y_0) = 0.$$

Regarding the parametrizations of the lemniscate in Eqs. (1.3) and (1.4), we have that

$$\alpha'(t) = \left(\frac{1 - 3t^4}{(1 + t^4)^2}, \frac{3t^2 - t^6}{(1 + t^4)^2} \right),$$

for every $t \in \mathbb{R} \setminus \{0\}$, is the velocity vector associated to each of the parametrizations at the point $\alpha_1(t)$. Therefore, for every $t \neq 0$ the tangent line at the point $\alpha_1(t)$ is parametrized by

$$\beta(s) = \alpha_1(t) + \alpha_1'(t)s = \frac{1}{(1 + t^4)^2} \left(t + t^5 + s - 3st^4, t^3 + t^7 + 3st^2 - st^6 \right),$$

for $s \in \mathbb{R}$. The normal line is given by

$$\gamma(s) = \frac{1}{(1 + t^4)^2} \left(t + t^5 - 3st^2 + st^6, t + t^5 + s - 3st^4 \right), \quad s \in \mathbb{R}.$$

The QR Code of Fig. C.2 links to an illustration of this situation.

The substitution $x_0 = \frac{t}{1+t^4}$ and $y0 = \frac{t^3}{1+t^4}$ shows that the two ways to describe the lines coincide.

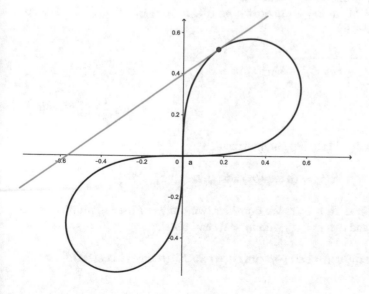

Fig. C.2 Tangent line in a point of a lemniscate. QR Code 22. Exercise 1.6

1.7. Searching for the points $(x, y) \in \mathbb{R}^2$ such that

$$d((x, y), (-a, 0)) \cdot d((x, y), (a, 0)) = a^2$$

we arrive at the expression

$$\sqrt{((x + a)^2 + y^2)((x - a)^2 + y^2)} = a^2,$$

which is equivalent to

$$(x^2 + 2ax + a^2 + y^2)(x^2 - 2ax + a^2 + y^2) = a^4,$$

$$(x^2 + a^2 + y^2)^2 - 4a^2x^2 = a^4,$$

and

$$(x^2 + y^2)^2 = 2a^2(x^2 - y^2).$$

1.8. The function $\alpha \circ \gamma^{-1} : I_2 \to \mathbb{R}^2$ is well-defined and belongs to $C^\infty(I_2)$ because it is the composition of regular functions. In addition to this, for every $t \in I_2$, one has that $(\alpha \circ \gamma)'(t) = (\gamma^{-1})'(t)\alpha'(\gamma^{-1}(t)) = \neq (0, 0)$ for every $t \in I_2$, because γ is a one-to-one function.

     Also, if $t_1, t_2 \in I_2$ with $\alpha \circ \gamma^{-1}(t_1) = \alpha \circ \gamma^{-1}(t_2)$, then $\gamma^{-1}(t_1) = \gamma^{-1}(t_2)$ because α is a one-to-one function. We conclude the result taking into account that γ is also a one-to-one function.

1.9. We apply (1.11) to the parametrization $((0, 2\pi), \alpha)$ given by (1.5). The arc length is given by

$$r \int_0^{2\pi} \sqrt{(1 - \cos(t))^2 + \sin^2(t)} dt = r \int_0^{2\pi} \sqrt{2 - 2\cos(t)} dt$$

$$= -4r \cos(\frac{t}{2})|_{t=0}^{t=2\pi} = 8r.$$

1.10. We apply Eq. (1.11) to the parametrization

$$\alpha(t) = (a \exp(bt) \cos(t), a \exp(bt) \sin(t))$$

for some $a, b \in \mathbb{R}$, $t \in \mathbb{R}$. We consider two points of the logarithmic spiral $\alpha(t_0) = x_0$ and $\alpha(t_1) = x_1$, with $t_0 < t_1$, and obtain

$$\alpha'(t) = (a \exp(bt)(b \cos(t) - a \sin(t)), a \exp(bt)(b \sin(t) + \cos(t))), \quad t \in \mathbb{R}.$$

Therefore,

$$\int_{t_0}^{t_1} \|\alpha'(t)\| \, dt = |a|\sqrt{b^2 + 1} \int_{t_0}^{t_1} e^{bt} dt = \frac{|a|}{b}\sqrt{b^2 + 1}(e^{bt_1} - e^{bt_0}).$$

1.11. We recall that the parametrization associated to the graph of $f \in C^\infty(I)$ is defined by $\alpha(t) = (t, f(t))$ for every $t \in I$. In view of Proposition 1.3.20, we have for all $t \in I$ that

$$\kappa_\alpha(t) = \frac{\det(\alpha'(t), \alpha''(t))}{\|\alpha'(t)\|^3} = \frac{\begin{vmatrix} 1 & f'(t) \\ 0 & f''(t) \end{vmatrix}}{(1 + f'(t))^{3/2}} = \frac{f''(t)}{(1 + f'(t))^{3/2}}.$$

1.12. Let $((0, 2\pi), \alpha)$ be the parametrization defined by $\alpha(t) = (a\cos(t), b\sin(t))$ for every $t \in (0, 2\pi)$. This parametrization draws the whole ellipse except for the point $(a, 0)$. A parametrization with the same expression can be defined for a regular curve contained in the ellipse which covers the point missing and the results can be extended to that point, so we omit further details. The curvature at the point $\alpha(t)$ is given by

$$\kappa_\alpha(t) = \frac{\det(\alpha'(t), \alpha''(t))}{\|\alpha'(t)\|^3} = \frac{\begin{vmatrix} -a\sin(t) & b\cos(t) \\ -a\cos(t) & -b\sin(t) \end{vmatrix}}{(a^2\sin^2(t) + b^2\cos^2(t))^{3/2}}$$

$$= \frac{ab}{(a^2\sin^2(t) + b^2\cos^2(t))^{3/2}}.$$

Observe that in the case where $a = b$ the curve turns into a circle with curvature given by the inverse of the radius, as stated in Example 1.3.19.

1.13. Let $(\mathbb{R}, \alpha)$ be the parametrization $\alpha(t) = (a\cosh(t), b\sinh(t))$. This parametrization draws one of the branches of the hyperbola. For the other, we consider the parametrization $(\mathbb{R}, \beta)$, with $\beta(t) = (-a\cosh(t), b\sinh(t))$. From now on, we work with $(\pm a\cosh(t), b\sinh(t))$. It holds that

$$\kappa_\alpha(t) = \frac{\det(\alpha'(t), \alpha''(t))}{\|\alpha'(t)\|^3} = \frac{\begin{vmatrix} a\sinh(t) & b\cosh(t) \\ a\cosh(t) & b\sinh(t) \end{vmatrix}}{(a^2\sinh^2(t) + b^2\cosh^2(t))^{3/2}}$$

$$= \frac{-ab}{(a^2\sinh^2(t) + b^2\cosh^2(t))^{3/2}}. \tag{C.1}$$

1.14. The matrix associated to the conic is

$$M = \begin{pmatrix} 1 & 1 & 0 \\ 1 & 1 & 1 \\ 0 & 1 & 1 \end{pmatrix}, \qquad M_0 = \begin{pmatrix} 1 & 1 \\ 1 & 1 \end{pmatrix}.$$

Fig. C.3 Orthogonal lines to
the symmetry axis of the
parabola. Exercise 1.14

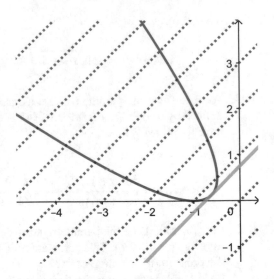

We have $\det(M) = -1$ and $\det(M_0) = 0$, and the conic is a parabola. It holds that $\lambda = 0$ is an eigenvalue of M_0, with $(1, -1)$ being an associated eigenvector. The axis of symmetry of the parabola is a line following the orthogonal direction to that vector. Therefore, we can find the axis of the parabola by cutting the parabola with a line of equation $y = x + c$, for a real parameter c. The vertex corresponds to the intersection of the only line whose intersection with the parabola is a single point (see Fig. C.3)

Inserting the parametric line $y = x + c$ in the equation of C we get the equation

$$4x^2 + (2 + 4c)x + 1 + c^2 = 0.$$

There is a single solution of this equation for the value $c = 3/4$, which determines the point $(-5/8, 1/8)$.

1.15. Given a reducible conic $C = \{(x, y) \in \mathbb{R}^2 : F(x, y) = 0\}$, then it holds that $F(x, y) = F_1(x, y)F_2(x, y)$, where none of the factors is a constant. Therefore, the degree of F_1 and F_2 is 1 for both of them, and each of them defines a line in $\mathbb{R}^2$. Now, we write $F_1(x, y) = a_1 x + b_1 y + c_1$ and $F_2(x, y) = a_2 x + b_2 y + c_2$, for some $a_j, b_j, c_j \in \mathbb{R}$, $j = 1, 2$. We distinguish the following situations:

- If the vector (a_1, b_1) is proportional to (a_2, b_2), then

  - if the vector (a_1, b_1, c_1) is proportional to (a_2, b_2, c_2), then the two lines coincide;
  - otherwise F defines a pair of parallel lines.

- otherwise, F defines a pair of secant lines.

1.16. Regarding the equation of the conic, we write the factorization

$$x^2 + 2y^2 + 3xy + 2x + 3y + 1 = (ax + by + c)(dx + ey + f).$$

However, the equations of the lines can be normalized to $x + by + c = 0$ and $x + ey + f = 0$ because $ad = 1$ so neither coefficient vanishes and one can divide both equations by them. After comparing the coefficients of the conic and $(x + by + c)(x + ey + f)$ we get that

$$be = 2, \quad e + b = 3, \quad f + c = 2, \quad bf + ce = 3, \quad \text{and } fc = 1.$$

From the last condition, $f = 1/c$ and from the third condition we deduce that $c = 1$. Therefore $f = 1$. An analogous reasoning can be applied to the first and second equations. There are two possibilities concerning (b, e), say $(2, 1)$ and $(1, 2)$. This corresponds to interchanging the roles of the two lines in the product. The remaining equation makes sense with respect to the previous one. We conclude that the conic is given by the lines of equations $x + 2y + 1 = 0$ and $x + y + 1 = 0$ (Fig. C.4).

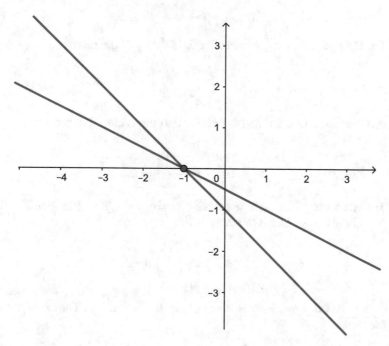

Fig. C.4 Two secant lines. Exercise 1.16

1.17. The matrices associated to the conic are

$$
M = \begin{pmatrix} 1 & 2 & -1 \\ 2 & 1 & 2 \\ -1 & 2 & 1 \end{pmatrix}, \quad M_0 = \begin{pmatrix} 1 & 2 \\ 2 & 1 \end{pmatrix}.
$$

We have $\det(M) = -16$ and $\det(M_0) = -3$ so the conic is a hyperbola. The solution of the system $M_0 c^T = b^T$, for $c = (c_1\ c_2)$ and $b = (-2\ 1)$, is given by $c = (4/3, -5/3)$, which determines the center of the hyperbola. The symmetry axes of the hyperbola are the lines at the center and with directions given by the eigenvectors of M_0. The eigenvalues of M_0 are the roots of $\det(M_0 - \lambda I) = 0$, where I stands for the identity matrix of order 2. The eigenvalue $\lambda_1 = 3$ is associated to the subspace generated by the vector $\omega_1 = (1, 1)$, so a first axis is the line

$$
(x, y) = c + \omega_1 t, \quad t \in \mathbb{R}.
$$

The second eigenvalue, $\lambda_2 = -1$, is associated to the eigenvector $\omega_2 = (1, -1)$. The second axis of the hyperbola is

$$
(x, y) = c + \omega_2 t, \quad t \in \mathbb{R}.
$$

The equation of the hyperbola after the rotation of the matrix

$$
\begin{pmatrix} \frac{\sqrt{2}}{2} & \frac{\sqrt{2}}{2} \\ \frac{\sqrt{2}}{2} & -\frac{\sqrt{2}}{2} \end{pmatrix}
$$

and the translation sending the center c to the origin of coordinates is given by

$$
3x^2 - y^2 + \frac{16}{3} = 0.
$$

The equations of the asymptotes are given by $y - 4/3 = m(x + 5/3)$, with m one of the two solutions of the equation

$$
(1\ m)\, M_0 \begin{pmatrix} 1 \\ m \end{pmatrix} = 0.
$$

This means that $m^2 + 4m + 1 = 0$, and $m = -2 \pm \sqrt{3}$. The asymptotes are the lines

$$
y + \frac{5}{3} = (-2 + \sqrt{3})(x - \frac{4}{3}), \text{ and}
$$

$$
y + \frac{5}{3} = (-2 - \sqrt{3})(x - \frac{4}{3}).
$$

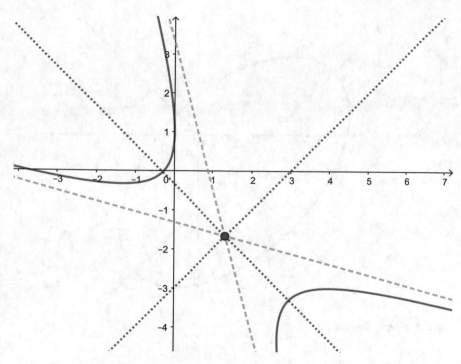

Fig. C.5 Hyperbola. Exercise 1.17

In order to determine the two possible hyperbolas having a common center, asymptotes and axes, we make a section through the hyperbola with one of the axes. We may obtain an empty intersection or two points. This completely determines the hyperbola, which is drawn in Fig. C.5.

1.18. The procedure followed in the previous exercise can be followed here as well. The eigenvalues of its associated matrix M_0 are $\lambda_1 = 3$ and $\lambda_2 = 1$, with $\det(M) = -5$ and $\det(M_0) = 3$. The conic is a real ellipse with center at $P = (-1/3, -1/3)$. We choose associated eigenvalues $\omega_1 = (1, 1)$ and $\omega_2 = (1, -1)$, which yield the axis of the ellipse parametrized by

$$(x, y) = c + t\omega_1, \quad \text{and } (x, y) = c + t\omega_2,$$

for $t \in \mathbb{R}$. The equation of the ellipse after the rotation of the matrix

$$\begin{pmatrix} \frac{\sqrt{2}}{2} & \frac{\sqrt{2}}{2} \\ \frac{\sqrt{2}}{2} & -\frac{\sqrt{2}}{2} \end{pmatrix}$$

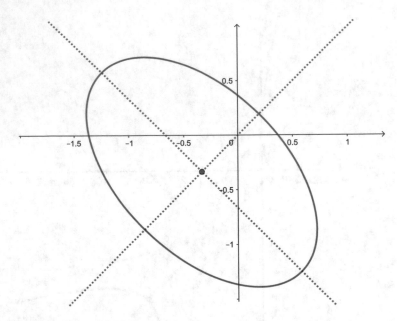

Fig. C.6 Ellipse. Exercise 1.18

and the translation sending the center c to the origin of coordinates is given by

$$3x^2 + y^2 - \frac{5}{3} = 0.$$

The ellipse is shown in Fig. C.6.

1.19. The first part follows by inserting the parametrization (1.25) in the equation of the conic. From the boundedness property of the parametrization, one realizes that the conix is an ellipse, without any further inspection.

Starting from the implicit equation of the conic and from the algorithm of parametrization of a conic described in the chapter, we obtain the following. The matrices associated to the conic are

$$M = \begin{pmatrix} -1 & 0 & 0 \\ 0 & 2 & 1 \\ 0 & 1 & 2 \end{pmatrix}, \qquad M_0 = \begin{pmatrix} 2 & 1 \\ 1 & 2 \end{pmatrix},$$

which is an ellipse, as noticed above. We have $\det(M)/\det(M_0) = -1$, and the eigenvalues associated to M_0 are $\lambda_1 = 1$ and $\lambda_2 = 3$. The equation of the ellipse becomes $3(x^\star)^2 + (y^\star)^2 - 1 = 0$ after an appropriate change of affine coordinates, consisting of the translation of the center of the ellipse,

$c = (0, 0)$, to the origin of coordinates (in this case the identity map, so no translation is needed) and the rotation of matrix

$$\begin{pmatrix} \frac{\sqrt{2}}{2} & \frac{\sqrt{2}}{2} \\ \frac{\sqrt{2}}{2} & -\frac{\sqrt{2}}{2} \end{pmatrix}.$$

In this coordinate system, the parametrization of the ellipse is

$$(x^\star, y^\star) = (\frac{\cos(t)}{\sqrt{3}}, \sin(t), \quad t \in \mathbb{R}.$$

The inverse change of coordinates yields the parametrization in the exercise.

1.20. We start from the parametrization of the lemniscate given in (1.8). Using the hint in the statements of the exercise, this parametrization can be written in the following form:

$$\begin{cases} x(t) = \frac{a\sqrt{2}\cos(t)}{\sin^2(t)+1} = \frac{-a\sqrt{2}(t^4-1)}{t^4+6t^2+1} \\ y(t) = \frac{a\sqrt{2}\cos(t)\sin(t)}{\sin^2(t)+1} = \frac{-2a\sqrt{2}(t^2-1)t}{t^4+6t^2+1} \end{cases} \quad t \in \mathbb{R},$$

We apply the technique used in Sect. 1.6 to obtain the implicit representation, by means of Theorem 1.6.2. We compute

$$\operatorname{res}_t (x(t^4 + 6t^2 + 1) + a\sqrt{2}(t^4 - 1), y(t^4 + 6t^2 + 1) + 2a\sqrt{2}(t^2 - 1)t$$

$$= 4a^4(32\sqrt{2}a^3x + 544a^2x^2 + 1280\sqrt{2}ax^3 - 88\sqrt{2}axy^2 + 1024x^4 + 1088x^2y^2 + y^4)$$

$$= -4096a^4(2a^2x^2 - 2a^2y^2 - x^4 - 2x^2y^2 - y^4). \qquad (C.2)$$

A simplification of the previous expression yields (1.7).

1.21. Let $y = m(x - 1)$ be a line passing at the point $(1, 0)$. Its intersection with the circle C is determined by the solution of the system

$$\begin{cases} x^2 + y^2 - 1 = 0 \\ y - mx + m = 0, \end{cases}$$

which has solutions in x determined by substitution $x^2 + (m(x - 1))^2 - 1 = 0$. This equation has $x = 1$ and $x = \frac{m^2-1}{m^2+1}$ as solutions. The first solution consists of the point $(1, 0)$, whilst the second one corresponds to the point $(\frac{m^2-1}{m^2+1}, -\frac{2m}{m^2+1})$. Therefore, a parametrization of the circle, except from the point $(1, 0)$ is given by $(\mathbb{R}, \beta)$, with

$$\beta(t) = \left(\frac{t^2 - 1}{t^2 + 1}, -\frac{2t}{t^2 + 1} \right), \quad \beta \in \mathbb{R}.$$

1.22. The line $y = m(x - a)$ crosses the ellipse at the solution of the system

$$\begin{cases} \frac{x^2}{a^2} + \frac{y^2}{b^2} - 1 = 0 \\ y - mx + ma = 0 \end{cases}$$

The substitution yields the intersection at the point $(a, 0)$ and also at $x = \frac{a^2((am)^2-b^2)}{(am)^2+b^2}$, so the other point at the intersection is $(\frac{a((am)^2-b^2)}{(am)^2+b^2}, \frac{2mab^2}{(am)^2+b^2})$, which determines the parametrization $(\mathbb{R}, \beta)$, with

$$\beta(t) = \left(\frac{a((at)^2 - b^2)}{(at)^2 + b^2}, \frac{2tab^2}{(at)^2 + b^2} \right), \quad \beta \in \mathbb{R}.$$

1.23. The line $y = m(x - a)$ crosses the hyperbola at the solution of the system

$$\begin{cases} \frac{x^2}{a^2} - \frac{y^2}{b^2} - 1 = 0 \\ y - mx + ma = 0, \end{cases}$$

which intersects the hyperbola at $(a, 0)$ and also at $x = \frac{a((am)^2+b^2)}{((am)^2-b^2)}$, which defines the point $(\frac{a((am)^2+b^2)}{((am)^2-b^2)}, \frac{2mab^2}{(am)^2-b^2})$ in the graph of the hyperbola. This corresponds to the parametrization $(\mathbb{R}, \beta)$, with

$$\beta(t) = \left(\frac{a((at)^2 + b^2)}{(at)^2 - b^2}, \frac{2tab^2}{(at)^2 - b^2} \right), \quad \beta \in \mathbb{R}.$$

Observe that the denominators in the previous expressions vanish at the value $t = b/a$ and $t = -b/a$, which correspond to the lines of equations $y = b/a(x - a)$ and $y = -b/a(x - a)$. These lines are parallel to the asymptotes of the hyperbola and only intersect the hyperbola at the point $(a, 0)$.

1.24. We follow the techniques stated in Sect. 1.7 to obtain the approximations required. For $N = 0$, we approximate the catenary $y(x) = \frac{a}{2}(\exp(\frac{x}{a}) + \exp(-\frac{x}{a}))$ by the constant polynomial

$$p_0(x) = y(1/2) = \frac{a}{2}(\exp(\frac{1}{2a}) + \exp(-\frac{1}{2a})).$$

In order to obtain the approximation for $N = 1$, we divide the segment $[0, 1]$ into three intervals of the same length, with end points at $x = 0, 1/3, 2/3, 1$. We evaluate $y(x)$ at $x = 1/3$ and $x = 2/3$ to get $y(1/3) = \frac{a}{2}(\exp(\frac{1}{3a})+\exp(-\frac{1}{3a}))$ and $y(2/3) = \frac{a}{2}(\exp(\frac{2}{3a})+\exp(-\frac{2}{3a}))$. The Lagrange

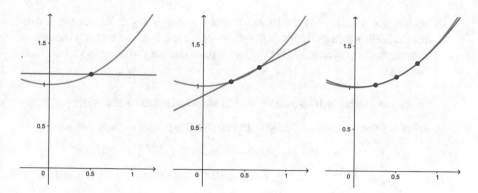

Fig. C.7 Catenary arc and three Lagrange approximations in [0, 1], $a = 1$. Exercise 1.24

interpolation polynomial is

$$p_1(x) = y(1/3)\frac{x - 2/3}{1/3 - 2/3} + y(2/3)\frac{x - 1/3}{2/3 - 1/3}$$

$$= \frac{a}{2}(-e^{\frac{-2}{3a}} - e^{\frac{2}{3a}} + 2e^{\frac{1}{3a}} + 2e^{\frac{-1}{3a}} + x(3e^{\frac{2}{3a}} - 3e^{\frac{1}{3a}} - 3e^{\frac{-1}{3a}} + 3e^{\frac{-2}{3a}})).$$

We have for $N = 3$ that the interval [0, 1] is divided by the points $1/4, 1/2, 3/4$. The Lagrange polynomial is defined by

$$p_2(x) = y(1/4)\frac{(x - 1/2)(x - 3/4)}{(1/4 - 1/2)(1/4 - 3/4)} + y(1/2)\frac{(x - 1/4)(x - 3/4)}{(1/2 - 1/4)(1/2 - 3/4)}$$

$$+ y(3/4)\frac{(x - 1/4)(x - 1/2)}{(3/4 - 1/4)(3/4 - 1/2)}$$

$$= \frac{a}{2}e^{-\frac{3}{4a}}\left((8e^{\frac{3}{2a}} - 16e^{\frac{5}{4a}} + 8e^{\frac{1}{a}} + 8e^{\frac{1}{2a}} - 16e^{\frac{1}{4a}} + 8)x^2\right.$$

$$+ 2(-3e^{\frac{3}{2a}} + 8e^{\frac{5}{4a}} - 5e^{\frac{1}{a}} - 5e^{-\frac{1}{2a}} + 8e^{\frac{1}{4a}} - 3)x + 3(-e^{\frac{5}{4a}} + e^{\frac{1}{a}} + e^{\frac{1}{2a}} - e^{\frac{1}{4a}}) + 1\left.\right).$$

Figure C.7 shows each of the three polynomials and the catenary in the interval $[-1, 1]$.

C.2   Chapter 2

2.1. One may consider the logarithmic spiral parametized by $(\mathbb{R}, \alpha)$, with $\alpha(t) = (a \exp(bt) \cos(t), a \exp(bt) \sin(t))$ (see (1.9)). A natural generalization of this curve to a space curve could be $(\mathbb{R}, \beta)$, with

$$\beta(t) = (a \exp(bt) \cos(t), a \exp(bt) \sin(t), t), \quad t \in \mathbb{R},$$

which coincides with the plane curve in plane view. Regarding Theorem 2.2.29, we have that it is sufficient to check the value of the torsion of the curve. Proposition 2.2.33 provides a practical way to compute the torsion. We have

$$\beta'(t) = (-a \exp(bt)(-b \cos(t) + \sin(t)), a \exp(bt)(b \sin(t) + \cos(t)), 1)$$

$$\beta''(t) = (-a \exp(bt)((1-b^2) \cos(t)+2b \sin(t)), a \exp(bt)((b^2-1) \sin(t)+2b \cos(t)), 0)$$

$$\beta'''(t) = (-a \exp(bt)((3b - b^3) \cos(t) + (3b^2 - 1) \sin(t)),$$

$$a \exp(bt)((b^3 - 3b) \sin(t) + (3b^2 - 1) \cos(t)), 0).$$

We conclude that

$$\tau_\beta(t) = \frac{1}{e^{2bt} a^2 + 1}, \quad t \in \mathbb{R},$$

which means that the curve is not a plane curve. The curve is contained in the surface of equation $z = \ln(\sqrt{x^2 + y^2})$. This concept is treated with in greater detail in Sect. 3.4. Figure C.8 shows the curve contained in the surface.

A similar approach can be followed with the so-called spiral of parametrization $(t \cos(t), t \sin(t))$, which gives rise to a well-known space curve, the **conical spiral of Pappus**, parametrized by $(\mathbb{R}, \beta)$, for

$$\beta(t) = (t \cos(t), t \sin(t), t), \quad t \in \mathbb{R},$$

which turns out to be a curve contained in a cone (see Fig. C.9)

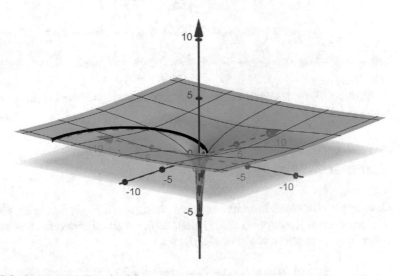

Fig. C.8 Space curve. Exercise 2.1

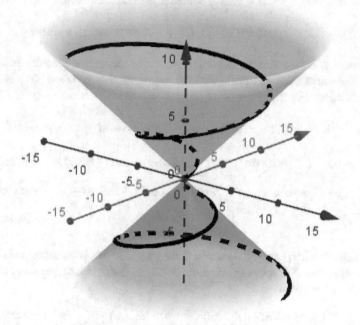

Fig. C.9 The conical spiral of Pappus. Exercise 2.1

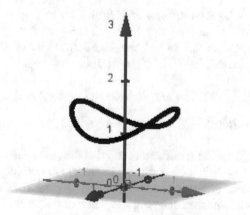

Fig. C.10 Example of space curve. Exercise 2.2

2.2. The projection of the curve onto the XY plane coincides with the unit circle, parametrized by $(\cos(t), \sin(t))$. The projection of the curve in the YZ plane defines the catenary-like curve $z = \frac{1}{2}(\exp(y) + \exp(-y))$, for $y \in [-1, 1]$. The projection of the curve in the XZ plane is the plane curve parametrized by $(\cos(t), \frac{1}{2}(\exp(\sin(t)) + \exp(-\sin(t))))$, $t \in \mathbb{R}$.

The curve is shown in Fig. C.10.

2.3. Let $R > r > 0$ and $n \in \mathbb{N}$. We consider the parametrization

$$\alpha(t) = ((R + r\cos(nt))\cos(t), (R + r\cos(nt))\sin(t), r\sin(nt)), \quad t \in \mathbb{R}.$$

First, we search for the inflection points of the curve, i.e., the points $\alpha(t)$ which satisfy the condition that the set of vectors $\{\alpha'(t), \alpha''(t)\}$ is linearly dependent. For every $t \in \mathbb{R}$ we have

$$\alpha'(t) = ((R + r\cos(nt))\cos(t), (R + r\cos(nt))\sin(t), r\sin(nt)),$$

$$\alpha''(t) = \Big(-r\cos(nt)n^2\cos(t) + 2r\sin(nt)n\sin(t) - \cos(t)R - \cos(t)r\cos(nt) ,$$

$$- r\cos(nt)n^2\sin(t) - 2r\sin(nt)n\cos(t) - \sin(t)R - \sin(t)r\cos(nt),$$

$$-r\sin(nt)n^2\Big).$$

Evaluating $\alpha'(t)$ and $\alpha''(t)$ at $t = 0$ we conclude that both vectors are linearly independent. We apply Proposition 2.2.21 to the parametrization of the space curve to arrive at

$$\kappa_\alpha(t) = \frac{(A_1 A_2 \cos(nt) + A_3 \cos(nt)^2 + 4Rr^3 cos(nt)^3 + (-r^4 n^2 + r^4)\cos(nt)^4)^{1/2}}{(R^2 + 2Rr\cos(nt) + r^2\cos(nt)^2 + r^2 n^2)^{(3/2)}},$$

with $A_1 = r^4 n^6 + R^4 + 4r^4 n^4 + r^2 n^4 R^2 + 4R^2 r^2 n^2$, $A_2 = 4R^3 r + 2R^3 rn^2 + 8Rr^3 n^2 + 4Rr^3 n^4$, and $A_3 = 6R^2 r^2 + 4r^4 n^2 - r^4 n^4 + 3r^2 n^2 R^2$.

From Proposition 2.2.33 and Theorem 2.2.29 we can verify that the curve is not plane. Indeed, $[\alpha'(t), \alpha''(t), \alpha'''(t)]$ is given by

$$- rn((r^2(n^2 - 1))\cos(nt)^3 + (-2rR(1 + 2n^2))\cos(nt)^2$$

$$+ (R^2(n^2 - 1) - 2r^2 n^4 - 4r^2 n^2)\cos(nt) + +Rrn^4 + 2Rrn^2).$$

Regarding Fig. C.11, one can verify that the curve can not be embedded in a plane for $n = 3$, $R = 3$ and $r = 1$. The curve considered is contained in a surface known as a torus and has interesting properties with respect to the tangent direction of the curve with respect to the surface.

2.4. The natural parametrization can be obtained by the change of the parameter provided in Proposition 2.2.11, i.e.,

$$\gamma^{-1}(t) = \int_{t_0}^{t} \|\alpha'(u)\| \, du.$$

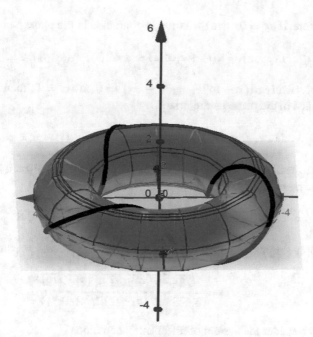

Fig. C.11 Example of a loxodrome contained in a torus. Exercise 2.3

The integrand is given by $\sqrt{1 + 4u^2}$, so the change of variable $2u = \sinh(s)$ leads to a much simpler integral. We conclude that

$$\gamma^{-1}(t) = \frac{1}{2}t\sqrt{1 + 4t^2} + \frac{1}{4}\text{arcsinh}(2t) - (\frac{1}{2}t_0\sqrt{1 + 4t_0^2} + \frac{1}{4}\text{arcsinh}(2t_0)).$$

2.5. For every $t \in \mathbb{R}$ we have

$$\alpha'(t) = (a_2, a_4 + 2a_5t, a_7 + 2a_8t + 3a_9t^2),$$

therefore, it defines a regular curve. We also have that

$$\alpha''(t) = (0, 2a_5, 2a_8 + 6a_9t), \quad \alpha'''(t) = (0, 0, 6a_9).$$

From Proposition 2.2.33, we get that it is a plane curve if and only if

$$[\alpha'(t), \alpha''(t), \alpha'''(t)] \equiv 0,$$

which is equivalent to $a_5a_9 = 0$. We observe from the proof of Theorem 2.2.29 that the plane containing the curve is the osculating plane.

Therefore, if $a_5 = 0$, then the curve is contained in the plane parametrized by

$$(x, y, z) = \alpha(t) + s_1\alpha'(t) + s_2\alpha''(t), \quad (s_1, s_2) \in \mathbb{R}^2,$$

or equivalently $a_4(x - a_1) + a_2(y - a_3) = 0$. If $a_9 = 0$, then the curve is contained in the plane parametrized by

$$(x, y, z) = \alpha(t) + s_1\alpha'(t) + s_2\alpha''(t), \quad (s_1, s_2) \in \mathbb{R}^2,$$

or equivalently $a_2(a_8a_4 - a_5a_7)(x - a_1) + a_2a_8(y - a_3) + a_2a_5(z - a_6) = 0$. If $a_5 = a_8 = 0$, then, we get the line

$$(x, y, z) = (a_1, a_3, a_6) + t(a_2, a_4, a_7), \quad t \in \mathbb{R}.$$

2.6. We apply Proposition 2.2.21 to the expressions above and get that

$$\kappa_\alpha(t) = \frac{2\sqrt{2 + 6t + 18t^2 + 18t^3 + 9t^4}}{(3 + 8t + 14t^2 + 12t^3 + 9t^4)^{3/2}}.$$

2.7. Let us consider the system of differential equations

$$\frac{dT_\alpha(t)}{dt} = cN_\alpha(t),$$

$$\frac{dN_\alpha(t)}{dt} = -cT_\alpha(t) + dB_\alpha(t)$$

$$\frac{dB_\alpha(t)}{dt} = -dN_\alpha(t)$$

with initial conditions given by $T_\alpha(0) = (1, 0, 0)$, $N_\alpha(0) = (0, 1, 0)$ and $B_\alpha(0) = (0, 0, 1)$. Its solution is such that $T_\alpha(t)$ equals

$$\left(\frac{c^2 \cos(\sqrt{c^2 + d^2}t) + d^2}{c^2 + d^2}, \frac{c \sin(\sqrt{c^2 + d^2}t)}{\sqrt{c^2 + d^2}}, \frac{cd(1 - \cos(\sqrt{c^2 + d^2}t))}{c^2 + d^2} \right).$$

This curve is a circle centered at $C = (\frac{d^2}{c^2+d^2}, 0, \frac{cd}{c^2+d^2})$ contained in the plane of equation $dx + cz = dc^2/(c^2 + d^2)$.

We assume $t_0 = 0$ without loss of generality, and have

$$\alpha(t) = \alpha(0) + \int_0^t T_\alpha(s)ds,$$

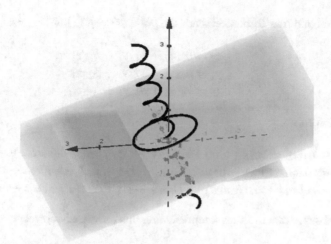

Fig. C.12 Construction of a curve of constant curvature and torsion. Exercise 2.7

i.e.,

$$\alpha(t) = \left(\frac{c^2 \sin(\sqrt{c^2 + d^2}t)}{(c^2 + d^2)^{3/2}} + \frac{d^2}{c^2 + d^2}t, \, -\frac{c \cos(\sqrt{c^2 + d^2}t)}{c^2 + d^2}, \right.$$

$$\left. \frac{cd}{c^2 + d^2}t - \frac{cd \sin(\sqrt{c^2 + d^2}t))}{(c^2 + d^2)^{3/2}} \right).$$

The above parametrization defines a helix. We display the previous geometric elements for $c = 3$ and $d = 1$ in Fig. C.12.

2.8. Let A_1 and A_2 be the matrices associated to these rigid transformations, of the form

$$A_1 = \begin{pmatrix} 1 & 0 & 0 & 0 \\ 0 & a_{11} & a_{12} & a_{13} \\ 0 & a_{21} & a_{22} & a_{23} \\ 0 & a_{31} & a_{32} & a_{33} \end{pmatrix}, \quad A_2 = \begin{pmatrix} 1 & 0 & 0 & 0 \\ 0 & b_{11} & b_{12} & b_{13} \\ 0 & b_{21} & b_{22} & b_{23} \\ 0 & b_{31} & b_{32} & b_{33} \end{pmatrix},$$

for some $a_{ij}, b_{ij} \in \mathbb{R}$. The first transformation sends a generic point $(x, y, z) \in \mathbb{R}^3$ to (x_1, y_1, z_1), with

$$\begin{cases} x_1 = a_{11}x + a_{12}y + a_{13}z \\ y_1 = a_{21}x + a_{22}y + a_{23}z \\ z_1 = a_{31}x + a_{32}y + a_{33}z \end{cases}$$

The second transformation sends the point (x_1, y_1, z_1) to (x_2, y_2, z_2), where

$$\begin{cases} x_2 = b_{11}x_1 + b_{12}y_1 + b_{13}z_1 \\ y_2 = b_{21}x_1 + b_{22}y_1 + b_{23}z_1 \\ z_2 = b_{31}x_1 + b_{32}y_1 + b_{33}z_1 \end{cases}$$

Therefore,

$$\begin{cases} x_2 = b_{11}(a_{11}x + a_{12}y + a_{13}z) + b_{12}(a_{21}x + a_{22}y + a_{23}z) + b_{13}(a_{31}x + a_{32}y + a_{33}z) \\ y_2 = b_{21}(a_{11}x + a_{12}y + a_{13}z) + b_{22}(a_{21}x + a_{22}y + a_{23}z) + b_{23}(a_{31}x + a_{32}y + a_{33}z) \\ z_2 = b_{31}(a_{11}x + a_{12}y + a_{13}z) + b_{32}(a_{21}x + a_{22}y + a_{23}z) + b_{33}(a_{31}x + a_{32}y + a_{33}z) \end{cases}$$

We observe that the matrix representing this movement corresponds to $A_3 = A_1 \cdot A_2$.

2.9. An argument analysis to that in the previous exercise can be followed, considering the inverse of the translation of vector (v_1, v_2, v_3), the translation of vector $-(v_1, v_2, v_3)$, the inverse of the rotation around a line of angle θ being the rotation around the same line of angle $-\theta$ and the reflection with respect to a plane being its own inverse.

2.10. First, we may assume without loss of generality that the plane is $z = 0$ after some rigid movement. The reflection of a point (x, y, z) is given by (x', y', z'), in (2.30). Then (x, y, z) equals (x', y', z') if and only if

$$\begin{pmatrix} 1 \\ x \\ y \\ z \end{pmatrix} = \begin{pmatrix} 1 & 0 & 0 & 0 \\ 0 & 1 & 0 & 0 \\ 0 & 0 & 1 & 0 \\ 0 & 0 & 0 & -1 \end{pmatrix} \begin{pmatrix} 1 \\ x \\ y \\ z \end{pmatrix},$$

which holds if and only if $z = 0$, i.e. if and only if the point belongs to the plane.

2.11. It holds that

$$\alpha'(t) = (-\sin(t), \cos(t), 1), \quad t \in \mathbb{R}.$$

with $\|\alpha'(t)\| = \sqrt{2}$. Therefore, $T_\alpha(t) = \frac{1}{\sqrt{2}}(-\sin(t), \cos(t), 1)$. In addition to this,

$$\alpha''(t) = (-\cos(t), -\sin(t), 0), \quad t \in \mathbb{R},$$

and $\alpha'(t) \times \alpha''(t) = (\sin(t), -\cos(t), 1)$, with $\|\alpha'(t) \times \alpha''(t)\| = \sqrt{2}$. From this, we deduce that $B_\alpha(t) = \frac{1}{\sqrt{2}}(\sin(t), -\cos(t), 1)$. We conclude with $N_\alpha(t) = B_\alpha(t) \times T_\alpha(t)$, i.e. $N_\alpha(t) = (-\cos(t), -\sin(t), 0)$.

2.12. We have

$$\alpha_1'(t) = (-\rho \sin(t), \rho \cos(t), 1), \quad t \in \mathbb{R},$$

with $\left\| \alpha_1'(t) \right\| = \sqrt{\rho^2 + 1}$,

$$\alpha_1''(t) = (-\rho \cos(t), -\rho \sin(t), 0), \quad t \in \mathbb{R}.$$

$$\alpha_1'''(t) = (\rho \sin(t), -\rho \cos(t), 0), \quad t \in \mathbb{R}.$$

We get that

$$\alpha_1'(t) \times \alpha_1''(t) = (\rho \sin(t), -\rho \cos(t), r^2), \quad t \in \mathbb{R},$$

with $\left\| \alpha_1'(t) \times \alpha_1''(t) \right\| = \rho \sqrt{1 + \rho^2}$. In light of Proposition 2.2.21, we obtain that

$$\kappa_{\alpha_1}(t) = \frac{\rho \sqrt{1 + \rho^2}}{(1 + \rho^2)^{3/2}} = \frac{\rho}{1 + \rho^2}.$$

On the other hand,

$$[\alpha_1'(t), \alpha_1''(t), \alpha_1'''(t)] = \rho^2,$$

From Proposition 2.2.33 we derive

$$\tau_{\alpha_1}(t) = \frac{1}{1 + \rho^2}.$$

2.13. Observe that this situation is more general that the one in the previous exercise, in which $h = 1$. Here, we compute

$$\alpha_2'(t) = (-\rho \sin(t), \rho \cos(t), h), \quad t \in \mathbb{R},$$

with $\left\| \alpha_2'(t) \right\| = \sqrt{\rho^2 + h^2}$,

$$\alpha_2''(t) = (-\rho \cos(t), -\rho \sin(t), 0), \quad t \in \mathbb{R}.$$

$$\alpha_2'''(t) = (\rho \sin(t), -\rho \cos(t), 0), \quad t \in \mathbb{R}.$$

We get that

$$\alpha_2'(t) \times \alpha_2''(t) = (h\rho \sin(t), -h\rho \cos(t), r^2), \quad t \in \mathbb{R},$$

with $\left\| \alpha_2'(t) \times \alpha_2''(t) \right\| = \rho\sqrt{h^2 + \rho^2}$. In light of Proposition 2.2.21, we obtain that

$$\kappa_{\alpha_2}(t) = \frac{\rho}{h^2 + \rho^2}.$$

On the other hand,

$$[\alpha_2'(t), \alpha_2''(t), \alpha_2'''(t)] = h\rho^2,$$

From Proposition 2.2.33 we derive

$$\tau_{\alpha_2}(t) = \frac{h}{h^2 + \rho^2}.$$

2.14. Computations analogous to those of the two previous exercises yield

$$\alpha_3'(t) = (-\rho\sin(t), \rho\cos(t), h'(t)), \quad t \in \mathbb{R},$$

with $\left\| \alpha_3'(t) \right\| = \sqrt{\rho^2 + (h'(t))^2}$,

$$\alpha_3''(t) = (-\rho\cos(t), -\rho\sin(t), h''(t)), \quad t \in \mathbb{R}.$$

$$\alpha_3'''(t) = (\rho\sin(t), -\rho\cos(t), h'''(t)), \quad t \in \mathbb{R}.$$

We get that

$$\alpha_3'(t) \times \alpha_3''(t) = (\rho(\cos(t)h''(t) + h'(t)\sin(t)),$$
$$- \rho(h'(t)\cos(t) - \sin(t)h''(t)), \rho^2), \quad t \in \mathbb{R},$$

with $\left\| \alpha_3'(t) \times \alpha_3''(t) \right\|$ being

$$\rho\sqrt{h'(t)^2 + h''(t)^2 + \rho^2}.$$

In light of Proposition 2.2.21, we obtain (2.37). On the other hand,

$$[\alpha_3'(t), \alpha_3''(t), \alpha_3'''(t)] = (h'(t) + h'''(t))\rho^2.$$

We derive (2.38) and (2.36).

We observe that the results obtained are coherent with the two previous exercises.

2.15. The parametrization is

$$\alpha_5(t) = (t\cos(t), t\sin(t), t), \quad t \in \mathbb{R}.$$

This parametrization corresponds to the conical spiral of Pappus, studied in the exercises of the previous chapter, whose graph is illustrated in Fig. C.9.

2.16. We have

$$\alpha_5'(t) = (\cos(t) - t\sin(t), \sin(t) + t\cos(t), 1), \quad t \in \mathbb{R},$$

with $\|\alpha_1'(t)\| = \sqrt{2 + t^2}$,

$$\alpha_5''(t) = (-2\sin(t) - t\cos(t), 2\cos(t) - t\sin(t), 0), \quad t \in \mathbb{R}.$$

$$\alpha_5'''(t) = (-3\cos(t) + t\sin(t), -3\sin(t) - t\cos(t), 0), \quad t \in \mathbb{R}.$$

We get that

$$\alpha_5'(t) \times \alpha_5''(t) = (-2\cos(t) + t\sin(t), -2\sin(t) - t\cos(t), 2 + t^2), \quad t \in \mathbb{R},$$

with $\|\alpha_5'(t) \times \alpha_5''(t)\| = t^4 + 8 + 5t^2$. From Proposition 2.2.21, we have

$$\kappa_{\alpha_5}(t) = \frac{t^4 + 8 + 5t^2}{(2 + t^2)^{3/2}}, \quad t \in \mathbb{R}.$$

Also,

$$[\alpha_5'(t), \alpha_5''(t), \alpha_5'''(t)] = 6 + t^2,$$

From Proposition 2.2.33

$$\tau_{\alpha_1}(t) = \frac{6 + t^2}{(t^4 + 8 + 5t^2)^2}, \quad t \in \mathbb{R}.$$

C.3   Chapter 3

3.1. Any point $P = (x, y, z) \in \mathbb{R}^3$ which belongs to the set of points is such that

$$\text{dist}(P, P_1)^2 + \text{dist}(P, P_2)^2 = 6,$$

or equivalently

$$(x + \sqrt{2})^2 + y^2 + z^2 + (x - \sqrt{2})^2 + y^2 + z^2 = 6.$$

The simplification of the previous expression yields $x^2 + y^2 + z^2 = 1$. Therefore, the set of points coincides with the sphere centered at the origin, and unit radius.

The substitution of the property in the statement of the problem yields the plane of equation $x = 3\sqrt{2}/2$.

3.2. It is straightforward to verify that the derivatives of every order of the function f can be obtained at every point $(x, y, z) \in \mathbb{R}^3$. In addition to this, we have

$$\frac{\partial^{j+k+\ell}}{\partial x^j \partial y^k \partial z^\ell} f(x, y, z) = (-1)^k \exp(x - y + z).$$

3.3. It straightforward to verify that the functions $p_1 \equiv 1$, $p_2 \equiv x$, $p_3 \equiv y$ and $p_4 \equiv z$ belong to $C^\infty(\mathbb{R}^3)$. Any other polynomial p can be written as a linear combination of finite products of the previous polynomials. This means that p is a regular function because it is the composition of regular functions in $\mathbb{R}^3$.

3.4. We consider the parametrization

$$X(u, v) = (v\cos(u), v\sin(u), v), \qquad v \in (0, h), u \in (0, 2\pi).$$

This parametrization does not consider a segment in the cone. However, this part can be obviated as it is a set of measure 0 in $\mathbb{R}^2$. We apply (3.3) and the Fubini theorem to arrive at

$$\int\int_{(0,2\pi)\times(0,h)} \left| \frac{\partial X}{\partial u}(u, v) \times \frac{\partial X}{\partial v}(u, v) \right| du\, dv$$

$$= \int_0^{2\pi} \int_0^h |(-v\sin(u), v\cos(u), 0) \times (\cos(u), \sin(u), 1)| dv\, du$$

$$= \int_0^{2\pi} \int_0^h |(v\cos(u), v\sin(u), -v)| dv\, du = 2\pi \int_0^h \sqrt{2} v\, dv = \sqrt{2}\pi h^2.$$

Compare this result with the formula of the area of a cone of generatrix of length $\sqrt{2}h$.

3.5. One can consider the parametrization of one hemisphere

$$X(u, v) = (u, v, \sqrt{1 - u^2 - v^2}), \qquad (u, v) \in D((0, 0), 1),$$

compute its area and multiply it by 2. This does not consider a circle contained in the sphere which is of measure zero, and the result remains unchanged. After applying Eq. (3.3), we find that the area is given by

$$2 \int\int_{D((0,0),1)} \frac{1}{\sqrt{1 - u^2 - v^2}} du\, dv.$$

The change of variables (u, v) to (t, θ) with $u = r\cos(\theta)$ and $v = r\sin(\theta)$ allows us to conclude that this area is given by

$$2 \int_0^{2\pi} \int_0^1 \frac{r}{\sqrt{1 - r^2}} dr d\theta = 4\pi.$$

Another approach to the exercise is to consider another coordinate system such as the spherical coordinate system (see Appendix A). The area is given in terms of the parametrization

$$X(\varphi, \theta) = (\sin(\theta)\cos(\varphi), \sin(\theta)\sin(\varphi), \cos(\theta)),$$

with $\varphi \in (0, 2\pi)$ and $\theta \in (0, \pi)$. The area is given by

$$\int\int_{(0,2\pi)\times(0,\pi)} \left| \frac{\partial X}{\partial \varphi}(\varphi, \theta) \times \frac{\partial X}{\partial \theta}(\varphi, \theta) \right| d\varphi d\theta$$

$$= \int_0^{2\pi} \int_0^{\pi} |(\sin^2(\theta)\cos(\varphi), \sin^2(\theta)\sin(\varphi), \sin(\theta)\cos(\varphi))| d\theta d\varphi$$

$$= 2\pi \int_0^{\pi} \sin(\theta)d\theta = 4\pi.$$

3.6. We start from the graph associated to the function $f : U \to \mathbb{R}^3$, where $U \subseteq \mathbb{R}^2$ is an open set. The graph is determined by the set $Gr(f) = \{(x, y, z) \in \mathbb{R}^3 : z - f(x, y) = 0\}$. In light of Proposition 3.1.20, we get that for every $P = (x_0, y_0, f(x_0, y_0))$ in the graph, a normal vector to $Gr(f)$ is

$$\frac{\left(-\frac{\partial f}{\partial x}(P), -\frac{\partial f}{\partial y}(P), 1\right)}{\sqrt{1 + \|\nabla f\|^2}}.$$

3.7. Observe that for any fixed $p > 0$, the surface is contained in the cube $[-1, 1]^3$. Indeed, the surface tends to become this cube for $p \to \infty$. The surface for $p = 2$ is the unit sphere.

We observe that the surface in the variables $|x|^{p/2}$, $|y|^{p/2}$ and $|z|^{p/2}$ is that of the unit sphere. Therefore, possible parametrizations are

$$(x, y, z) = (\pm(\cos(u)\sin(v))^{2/p}, \pm(\sin(u)\sin(v))^{2/p}, \pm\cos(v)^{2/p}),$$

for $u \in (0, 2\pi)$ and $v \in (0, \pi)$, combining the signs $+$ and $-$.

Figure C.13 illustrates the case $p = 1/2$.

3.8. It is straightforward to verify the statement of the exercise by plugging the three expressions of the components of the parametrization of the curve into the implicit expressions of the surfaces. Both surfaces are quadrics. The equation

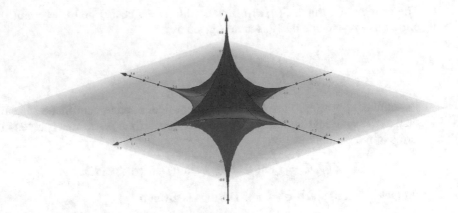

Fig. C.13 Lamé surface for $p = 1/2$. Exercise 3.7

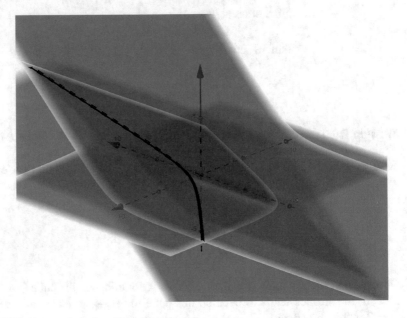

Fig. C.14 Intersection of surfaces. Exercise 3.8

defining the first one can be written as $x = z - 1/z$, so it is a hyperbolic cylinder along the OY axis. The second surface is related to the function $y = z + 1/z$, so it is a hyperbolic cylinder along the OX axis. The parametrization of the curve satisfies the condition that $x(t) + y(t) = 2z(t)$, which means that the curve is contained in the plane of equation $x + y - 2z = 0$. In addition to this, we can verify that the curve is part of a hyperbola due to the fact that it is the intersection of a plane with a hyperbolic cylinder. Figure C.14 illustrates the geometric elements involved in the exercise.

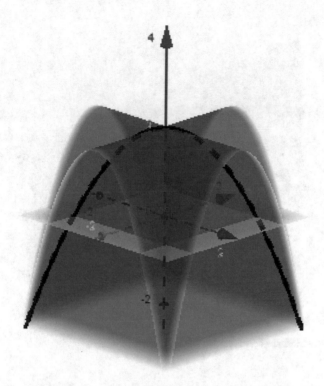

Fig. C.15 Intersection of two parabolas. Exercise 3.9

3.9. Assume the two parabolic cylinders are defined by the implicit equations $z = -ax^2 + b$, and $z = -ay^2 + b$, for some fixed $a, b > 0$. It turns out that the intersection is a curve parametrized by

$$t \mapsto (t, t, -at^2 + b), \quad t \in \mathbb{R}.$$

This is a parabola contained in the plane $x - y = 0$. Figure C.15 shows the situation in the case $a = 1, b = 2$.

C.4   Chapter 4

4.1. The line r has directing vector w given by

$$w = \begin{vmatrix} i & j & k \\ 1 & 2 & 0 \\ 1 & 1 & 1 \end{vmatrix} = (2, -1, -1).$$

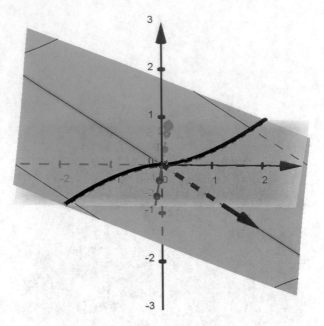

Fig. C.16 Cylindrical surface. Exercise 4.1

A regular parametrization of the curve d is $(\mathbb{R}, \alpha)$, with $\alpha(t) = (t, t^{5/3}, 0)$, for all $t \in \mathbb{R}$. Therefore, the cylindrical surface is parametrized by $(\mathbb{R}^2, X)$, with

$$X(u, v) = \alpha(u) + v * w = (u + 2v, u^{5/3} - v, -v), \quad (u, v) \in \mathbb{R}^2.$$

Figure C.16 shows the configuration of the construction.

4.2. The directrix d can be parametrized by $(\mathbb{R}, \alpha)$, with $\alpha(t) = (3\cos(t), 2\sin(t), 3)$, for $t \in \mathbb{R}$. The conical surface is parametrized by $(\mathbb{R}^2, X)$, with

$$X(u, v) = (3\cos(u), 2\sin(u), 3) + v(3\cos(u) - 1, 2\sin(u) - 1, 3 + (3 - 1)v)$$

$$= (3\cos(u) + v(3\cos(u) - 1), 2\sin(u) + v(2\sin(u) - 1), 3 + 2v), \quad u, v \in \mathbb{R}.$$

The elements involved in the construction appear in Fig. C.17.

The implicit equation of the surface in this simple case can be directly obtained by solving the system

$$\begin{cases} 3\cos(u) + v(3\cos(u) - 1) = x \\ 2\sin(u) + v(2\sin(u) - 1) = y \\ 3 + 2v = z \end{cases}$$

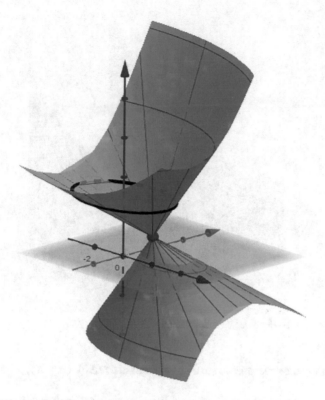

Fig. C.17 Conical surface. Exercise 4.2

in the parameters. We observe that the following relation holds for all (u, v) with $v \neq -1$:

$$\frac{x+v}{3+3v} = \cos(u), \quad \frac{y+v}{2+2v} = \sin(u).$$

Therefore,

$$\left(\frac{x+v}{3+3v}\right)^2 + \left(\frac{y+v}{2+2v}\right)^2 = 1.$$

Also, one can substitute in the previous equality $v = \frac{z-3}{2}$, which yields the equation

$$\frac{1}{9}(2x+z-3)^2 + \frac{1}{4}(2y+z-3)^2 = (2+z-3)^2,$$

which extends the previous relation for all values of the parameters.

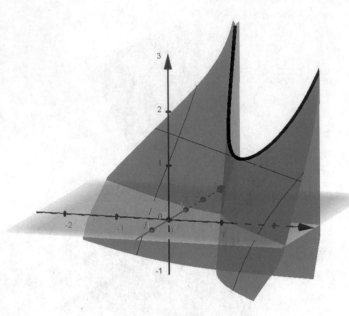

Fig. C.18 Tangent developable surface. Exercise 4.3

4.3. The tangent developable surface is parametrized by $(\mathbb{R}^2, X)$, with

$$X(u, v) = (\exp(u), \exp(-u), \exp(u^2)) + v(\exp(u), -\exp(-u), 2u \exp(u^2)),$$

$$= (\exp(u)(v + 1), (1 - v) \exp(-u), (1 + 2uv) \exp(u^2)), \quad (u, v) \in \mathbb{R}^2.$$

The surface is shown in Fig. C.18.

4.4. The matrix of the surface is

$$M = \begin{pmatrix} 0 & 0 & 0 & 0 \\ 0 & -2 & 0 & 0 \\ 0 & 0 & 0 & 1 \\ 0 & 0 & 1 & 0 \end{pmatrix}.$$

We have $\det(M) = 0$ and $\det(M_0) = 2 \neq 0$. The eigenvalues associated to the matrix M_0 are $\{-1, 1, -2\}$. Therefore, the quadric is a cone. By plugging the parametrization $(\mathbb{R}^2, X)$ into the equation of the cone, we find that it is indeed a parametrization of the cone. Observe that, from the procedure to parametrize conical surfaces, it holds that the surface is a cone with parabolic directrix. Figure C.19 illustrates the exercise.

4.5. It is straightforward to verify that

$$F(x, y, z) = z - (1 - \frac{y}{4}) \sin(\frac{x}{3}) - \frac{y}{7}$$

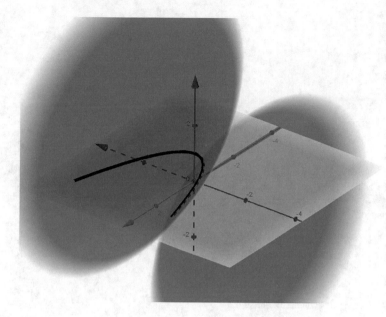

Fig. C.19 Quadric. Exercise 4.4

describes the previous parametrization implicitly. Regarding Proposition 3.1.20, the tangent plane at the point P is determined by

$$\frac{\partial F}{\partial x}(P) = -(1 - \frac{y}{4})\cos(\frac{x}{3})\frac{1}{3}|_P = -\frac{1}{4}.$$

$$\frac{\partial F}{\partial y}(P) = \frac{1}{4}\sin(\frac{x}{3}) - \frac{1}{7}|_P = -\frac{1}{7}.$$

$$\frac{\partial F}{\partial z}(P) = 1.$$

Therefore, the tangent plane is given by

$$-\frac{1}{4}(x - 3\pi) - \frac{1}{7}(y - 7) + z - 1 = 0.$$

Figure C.20 illustrates the situation. In order to check that the segment is contained in the surface, it suffices to insert the expressions of x, y and z in the implicit expression of the surface and check that it holds for every $t \in \mathbb{R}$.

Fig. C.20 Geometric scheme. Exercise 4.5

4.6. The matrices associated to the quadric are

$$M = \begin{pmatrix} 1 & 1 & -1 & 0 \\ 1 & 1 & 0 & 0 \\ -1 & 0 & 1 & 2 \\ 0 & 0 & 2 & 1 \end{pmatrix}, \quad M_0 = \begin{pmatrix} 1 & 0 & 0 \\ 0 & 1 & 2 \\ 0 & 2 & 1 \end{pmatrix}.$$

We have the determinant of M is -1, and the determinant of M_0 is -3. In addition to this, the eigenvalues associated to M_0 are $\lambda_1 = 3$, $\lambda_2 = -1$ and $\lambda_3 = 1$, which entails that the quadric is a hyperboloid of two sheets. The center of the hyperboloid is located at $C = (c_1, c_2, c_3)$, where

$$\begin{pmatrix} 1 & 0 & 0 \\ 0 & 1 & 2 \\ 0 & 2 & 1 \end{pmatrix} \begin{pmatrix} c_1 \\ c_2 \\ c_3 \end{pmatrix} = \begin{pmatrix} -1 \\ 1 \\ 0 \end{pmatrix}.$$

Therefore, the solution of the system provides $C = (-1, -1/3, 2/3)$. The axes of the quadric are determined by the lines at C and direction given by eigenvectors associated to each eigenvalue. It is straightforward to verify that $\omega_1 = (0, 1, 1)$, $\omega_2 = (0, -1, 1)$ and $\omega_3 = (1, 0, 0)$ are eigenvectors associated to λ_1, λ_2 and λ_3, respectively. The axis of the hyperboloid are

$$(x, y, z) = C + \omega_1 t, \quad (x, y, z) = C + \omega_2 t, \quad (x, y, z) = C + \omega_3 t,$$

for $t \in \mathbb{R}$. After the translation sending the origin of coordinates to C and the transformation $\varphi : \mathbb{R}^3 \to \mathbb{R}^3$ of matrix

$$
\begin{pmatrix}
0 & 0 & 1 \\
\frac{\sqrt{2}}{2} & -\frac{\sqrt{2}}{2} & 0 \\
\frac{\sqrt{2}}{2} & \frac{\sqrt{2}}{2} & 0
\end{pmatrix},
$$

the equation of the quadric in this new coordinate system is

$$
3(x^{\star\star})^2 - (y^{\star\star})^2 + (z^{\star\star})^2 + \frac{1}{3} = 0.
$$

Figure C.21 illustrates the quadric, its center, and its axes.

4.7. The matrices associated to the quadric are

$$
M = \begin{pmatrix}
-1 & 0 & -2 & 0 \\
0 & 1 & 0 & 0 \\
-2 & 0 & 2 & 0 \\
0 & 0 & 0 & 3
\end{pmatrix}, \qquad
M_0 = \begin{pmatrix}
1 & 0 & 0 \\
0 & 2 & 0 \\
0 & 0 & 3
\end{pmatrix}.
$$

Fig. C.21 Geometric scheme. Exercise 4.6

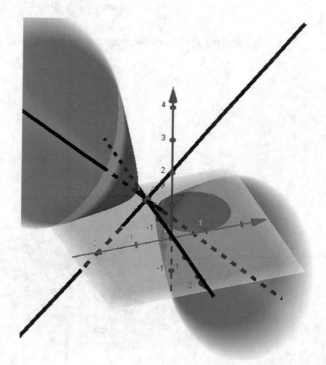

We have that the determinant of M is -18, and the determinant of M_0 is 6. In addition to this, the eigenvalues associated to M_0 are $\lambda_1 = 1$, $\lambda_2 = 2$ and $\lambda_3 = 3$, which means that the quadric is an ellipsoid. The center of the hyperboloid is located at $C = (c_1, c_2, c_3)$, where

$$\begin{pmatrix} 1 & 0 & 0 \\ 0 & 2 & 0 \\ 0 & 0 & 3 \end{pmatrix} \begin{pmatrix} c_1 \\ c_2 \\ c_3 \end{pmatrix} = \begin{pmatrix} 0 \\ 2 \\ 0 \end{pmatrix}.$$

Therefore, the solution of the system provides $C = (0, 1, 0)$. The axes of the quadric are determined by the lines at C and direction given by eigenvectors associated to each eigenvalue. It is straightforward to verify that $\omega_1 = (1, 0, 0)$, $\omega_2 = (0, 1, 0)$ and $\omega_3 = (0, 0, 1)$ are eigenvectors associated to λ_1, λ_2 and λ_3, respectively. The axis of the ellipsoid are

$$(x, y, z) = C + \omega_1 t, \quad (x, y, z) = C + \omega_2 t, \quad (x, y, z) = C + \omega_3 t,$$

for $t \in \mathbb{R}$. After the translation sending the origin of coordinates to C the equation of the quadric in this new coordinate system is

$$(x^{**})^2 + 2(y^{**})^2 + 3(z^{**})^2 - 3 = 0.$$

Figure C.22 illustrates the quadric, its center, and its axis.

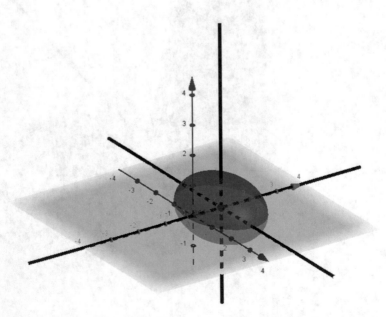

Fig. C.22 Geometric scheme. Exercise 4.7

4.8. The matrices associated to the quadric are

$$M = \begin{pmatrix} -1 & 0 & 0 & 0 \\ 0 & 1 & m & 0 \\ 0 & m & 1 & 0 \\ 0 & 0 & 0 & 1 \end{pmatrix}, \quad M_0 = \begin{pmatrix} 1 & m & 0 \\ m & 1 & 0 \\ 0 & 0 & 1 \end{pmatrix}.$$

We have that the determinant of M is $m^2 - 1$, and the determinant of M_0 is $1 - m^2$. The eigenvalues of M_0 are $\lambda_1 = 1$, $\lambda_2 = 1 - m$ and $\lambda_3 = 1 + m$. Therefore, we distinguish different cases:

- If $m \neq 1$, and $m \neq -1$ then none of the eigenvalues of M_0 are zero.

  - If $m < -1$, the sign of the eigenvalues determine a hyperboloid of one sheet.
  - If $-1 < m < 1$, the sign of the eigenvalues determine an ellipsoid. If $m = 0$, it turns out to be the unit sphere.
  - If $m > 1$, then the quadric is again a hyperboloid of one sheet.

- If $m = -1$, the equation of the quadric is $x^2 + y^2 + z^2 + 2xy - 1 = 0$, or equivalently $(x + y)^2 + z^2 = 1$, which determines a cylinder with circular directrix, with the center of the circles at the points of the line $x + y = 0, z = 0$.
- If $m = 1$, the equation of the quadric is $x^2 + y^2 + z^2 + 2xy - 1 = 0$, or equivalently $(x - y)^2 + z^2 = 1$, which determines a cylinder with circular directrix, with the center of the circles at the points of the line $x - y = 0, z = 0$.

4.9. The procedure rests on the parametric representation of a surface constructed by the segments joining two points moving in two curves, as described in Sect. 4.3. Let (I, α_1) be $I = (0, 2\pi)$ and $\alpha_1(t) = (\cos(t), \sin(t) - 1/2, 0)$, for every $t \in (0, 2\pi)$. This corresponds to the parametrization of a circle of unit radius located at the floor plane, and center at the point $(0, -1/2, 0)$. We also consider the parametrization (I, α_2), where $\alpha_2(t) = (0, \cos(u) + 1/2, \sin(u))$, for every $u \in I$, i.e., the curve (I, α_2) corresponds to a circle of unit radius at the plane $x = 0$ and centered at $(0, 1/2, 0)$. The points of the oloid are obtained by joining them.

   In the case of the sphericon, we consider the half-circles parametrized by (I_1, β_1) and (I_2, β_2), with $I_1 = (-\pi, 0)$ and $I_2 = (-\pi/2, \pi/2)$, with $\beta_1(t) = (\cos(t), \sin(t), 0)$, and $\beta_2(t) = (0, \cos(t), \sin(t))$. The segments joining both half-circles determine the sphericon within the four parametrizations considered.

4.10. We proceed to check whether the mean curvature (see Definition 4.7.3) of the helicoid in the statements vanishes. For this purpose, we compute the first and second order partial derivatives of $X(u, v)$, the normal vector given by the cross product of the partial derivatives, and the first and second fundamental forms associated.

We find that, at a vector $\omega = (\omega_1, \omega_2) \neq (0, 0)$ associated to the tangent plane of the associated surface at a point $P = X(u, v)$, the first fundamental form is given by

$$I(\omega) = \omega_1^2 + \omega_2^2(1 + u^2),$$

and

$$II(\omega) = -\frac{2\omega_1\omega_2}{u^2 + 1}.$$

Therefore, the curvature at vector ω is

$$k_n(\omega) = -\frac{2\omega_1\omega_2}{(u^2 + 1)((u^2 + 1)\omega_2^2 + \omega_1^2)}.$$

The above is a function of two variables with singular points at $W = (\omega_1, \pm\sqrt{u^2 + 1})$. We observe that in polar coordinates

$$(\omega_1, \omega_2) = (\rho\cos(\theta), \rho\sin(\theta)),$$

the function k_n is defined by

$$k_n(\rho, \theta) = -\frac{2\cos(\theta)\sin(\theta)}{(u^2 + 1)(u^2\sin(\theta) + 1)}.$$

The function

$$f(\theta) = \frac{\cos(\theta)\sin(\theta)}{u^2\sin^2(\theta) + 1}$$

attains its absolute maximum at $\theta_1 = \arccos(\sqrt{\frac{u^2+1}{u^2+2}})$ and absolute minimum at $\theta_2 = \pi - \arccos(\sqrt{\frac{u^2+1}{u^2+2}})$, with $f(\theta_1) = \frac{1}{2\sqrt{u^2+1}}$ and $f(\theta_2) = -\frac{1}{2\sqrt{u^2+1}}$, which means that the maximum of k_n is opposite the minimum quantity of k_n. Therefore, the mean curvature at each point of the surface is null, and the helicoid is a minimal surface.

4.11. Instead of following the construction at the beginning of the chapter, we make use of Rodrigues' rotation formula (see Theorem 4.4.1). We consider the ellipse of equations

$$\frac{(x - R)^2}{a^2} + \frac{z^2}{b^2} = 1, \quad y = 0,$$

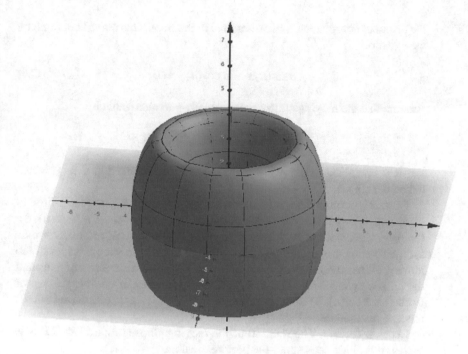

Fig. C.23 Elliptic torus. Exercise 4.11

with center at $(R, 0, 0)$ and parametrized by

$$(R + a\cos(v), 0, b\sin(v)), \quad v \in \mathbb{R}.$$

The rotation of this ellipse around the OZ axis determines the surface parametrized by

$$\begin{cases} x = R\cos(u) + a\cos(v)\cos(u) \\ y = R\sin(u) + a\cos(v)\sin(u) \\ z = \quad\quad b\sin(v) \end{cases} \tag{C.3}$$

Note that this structure is similar to that of the torus (3.9).

Figure C.23 illustrates the case $a = 1$ and $b = 3$.

4.12. There might be several different interpretations of the geometric object to be constructed. We will assume that the origin lies inside the plane in which the moving ellipse is contained. Other interpretations could consider othogonality with respect to the ellipse at the plan floor.

For every point fixed in the ellipse at the floor, say $(a\cos(u), b\sin(u), 0)$, for some $a, b > 0$ which define the ellipse, we consider the plane at that point, the origin of coordinates, and orthogonal to the floor. This plane is defined by the equation

$$b \sin(u)x - a \cos(u)y = 0. \tag{C.4}$$

Assume that the moving ellipse to be initially parametrizable by

$$(c \cos(v), 0, d \sin(v)), \quad v \in \mathbb{R}, \tag{C.5}$$

i.e., it is contained in the plane $y = 0$, and for some fixed $c, d > 0$ which define the ellipse.

From this point of departure, we proceed with a translation of the ellipse in (C.5) which puts the center of the ellipse at the point $(a \cos(u), b \sin(u), 0)$, for every choice of $t \in \mathbb{R}$. Afterwards, we apply Rodrigues' rotation formula for the rotation around the OZ axis with the angle t, in such a way that the moving ellipse has completed a rotation around itself after a rotation around the ellipse at the floor. The first translation consists in adding the vector $(a \cos(u), b \sin(u), 0)$ to the parametrization of the rotated ellipse, which we proceed to compute.

Let $\theta \in (0, 2\pi)$. The rotation of the moving ellipse around the OZ axis such at this angle determines the parametrization

$$(c \cos(v) \cos(\theta), c \cos(v) \sin(\theta), d \sin(v).$$

Now, taking into account that the ellipse is contained in the plane (C.4), we get the following condition on θ: $b \sin(u) \cos(\theta) = a \cos(u) \sin(\theta)$, i.e.,

$$\theta = \arctan(\frac{b}{a} \tan(u)),$$

for almost all values of u. The parametrization of the ellipse after this reasoning is

$$\begin{cases} x = a \cos(u) + c \cos(v) \cos(\arctan(\frac{b}{a} \tan(u))) \\ y = b \sin(u) + c \cos(v) \sin(\arctan(\frac{b}{a} \tan(u))) \\ z = \qquad d \sin(v) \end{cases}$$

for u, v. Observe that the case $a = b = R$ coincides with (C.3).

Figure C.24 illustrates the particular case in which $a = d = 2$ and $c = 1$, $b = 3$. It is also worth mentioning that in "extreme" situations, the surface becomes deformed. For example, draw the surface for $b = 3$, and $a = c = d = 2$ and find out what happens with the hole of the torus.

4.13. These kind of surfaces have been of great importance when constructing conoid shell structures in architecture (Dolezal 2011) (Fig. C.25).

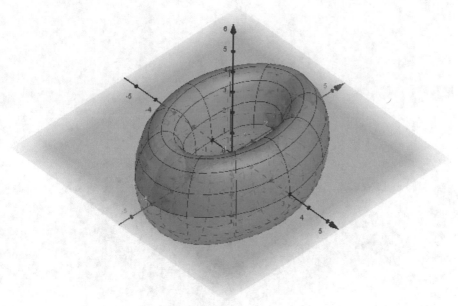

Fig. C.24 Generalized elliptic torus. Exercise 4.12

Fig. C.25 Conoid structure. Exercise 4.13

Let $h > 0$. For every $u \in (0, \pi)$, the point $(\cos(u), h, \sin(u))$ is associated to the point $(\cos(u), 0, 0)$ in L. Therefore, the conoid structure is parametrized by

$$(x, y, z) = \frac{h - v}{h}(\cos(u), 0, 0) + \frac{v}{h}(\cos(u), h, \sin(u)),$$

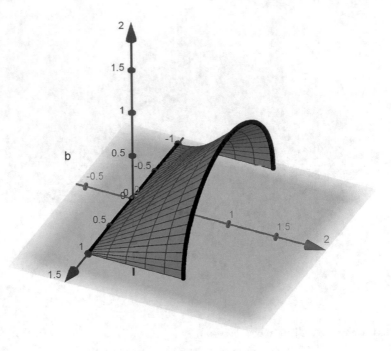

Fig. C.26 Second conoid structure. Exercise 4.13

for $0 < u < \pi$ and $v \in (0, h)$, which turns out to be rewritten in the form

$$(x, y, z) = (\cos(u), v, \frac{v}{h}\sin(u)).$$

Figure C.26 illustrates the construction for the case $h = 1$.

We now give an implicit description of the surface. Taking into account the classic trigonometric formula, we derive that $x^2 + h/vz^2 = 1$ for all the points in the surface. Moreover, $y = v$, and the surface is contained in the following

$$S = \{(x, y, z) \in \mathbb{R}^3 : x^2y^2 + hz^2 - y^2 = 0\}.$$

4.14. We apply Rodrigues' formula for the revolution around the OZ axis of the tractrix. We find that the surface of revolution is parametrized by

$$X(u, v) = \left(\frac{2}{e^u + e^{-u}} \cos(v), \frac{2}{e^u + e^{-u}} \sin(v), u - \frac{e^{2u} - 1}{e^{2u} + 1} \right),$$

for $u, v \in \mathbb{R}$.

The pseudosphere is represented in Fig. C.27.

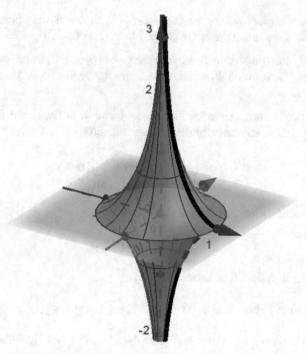

Fig. C.27 Pseudosphere. Exercise 4.14

In order to obtain the curvature at every point, we proceed as in the third exercise, to get that

$$k_n(\omega_1, \omega_2) = -\frac{\sinh(u)(\omega_1^2 - \omega_2^2)}{\sinh(u)^2\omega_1^2 + \omega_2^2}.$$

The nullity of the two partial derivatives of k_n determines two directions, corresponding to the canonical basis of $\mathbb{R}^2$. We get that the maximum of k_n is attained for the direction $(0, 1)$, and $k_n(0, 1) = \sinh(u)$. On the other hand, the minimum is attained at $k_n(1, 0) = -\frac{1}{\sinh(u)}$.

Therefore, the mean curvature is given by $\frac{\cosh^2(u)-2}{\sinh(u)}$, and the Gaussian curvature is constant and negative for all (u, v). Indeed, its value is always -1.

4.15. We write $\alpha = (\alpha_1, \alpha_2, C)$ and $\beta(s) = (\beta_1(s), \beta_2(s))$. Let $t \in I$ fixed, with $\alpha(t)$ in $\alpha(I)$. The plane at $\alpha(t)$ with normal vector being $\alpha'(t)$ has equation

$$\alpha_1'(t)(x - \alpha_1(t)) + \alpha_2'(t)(y - \alpha_2(t)) = 0.$$

Assume the plane where the curve $\beta_t(I_t)$ is drawn is the plane $z = 0$. The next two steps move the curve (I_t, β_t) to the appropriate position:

- Apply the translation T_t moving the point $(\beta_t(s_t), 0)$ to the point $\alpha(t)$.
- Apply the rotation Φ_t which transforms the vector $(0, 0, 1)$ into the vector $T_\alpha(t)$.

These two motions describe the correct location of the curve $\tilde{\beta}_t(I_t)$. Therefore, the parametrization of the surface is given by (U, X), with

$$U = \bigcup_{t \in I} \bigcup_{s \in I_t} \{(t, s)\} = \bigcup_{t \in I} \{t\} \times I_t,$$

and for all $(t, s) \in U$, with $s \in I_t$, then

$$X(t, s) = \Phi_t \circ T_t(\beta_t(s), 0).$$

The translation T_t is defined by

$$T_t(x, y, z) = (x + \alpha_1(t) - \beta_1(s_t), y + \alpha_2(t) - \beta_2(s_t), z + \alpha_3(t)).$$

The rotation Φ_t transforms the plane $z = 0$ into the plane $\alpha_1'(t)x + \alpha_2'(t)y = 0$. It is determined as follows. The tangent vector at t is given by

$$\left(\frac{\alpha_1'(t)}{\sqrt{(\alpha_1')^2(t) + \alpha_2'(t)}}, \frac{\alpha_2'(t)}{\sqrt{(\alpha_1')^2(t) + \alpha_2'(t)}}, 0 \right).$$

The curve after rotation is parametrized by

$$\left(\beta_1(s) \frac{\alpha_2'(t)}{\sqrt{(\alpha_1')^2(t) + \alpha_2'(t)}}, -\beta_1(s) \frac{\alpha_1'(t)}{\sqrt{(\alpha_1')^2(t) + \alpha_2'(t)}}, \beta_2(s) \right),$$

for $s \in I_t$.

We apply this procedure to the following example: Let (I, α) be the curve determined by $I = (-10, 10)$ and

$$\alpha(t) = (t, \cos(t), 1), \quad t \in I.$$

For every $t \in I$ we take $(I_t, \beta_t) = ((-1, 1), \beta)$, with

$$\beta(s) = (s, -s^2), \quad s \in (-1, 1),$$

and $\beta_t(s_t) = (0, 0)$ for every $t \in I$, i.e. $s_t = 0$ for all $t \in I$.

The translation is $T_t(x, y, z) = (x + \alpha_1(t), y + \alpha_2(t), z + \alpha_3(t))$, and the rotation Φ_t draws the curve

$$s \mapsto \left(s \frac{-\sin(t)}{\sqrt{1 + \sin^2(t)}}, -s \frac{1}{\sqrt{1 + \sin^2(t)}}, -s^2 \right), \quad s \in (-1, 1).$$

Figure C.29 illustrates the situation. The QR code in Fig. C.28 links to the construction of the surface by means of the sweeping curve.

A more direct construction of a surface of a similar nature is to consider the curve described by β as a parametrization of a space curve and consider

$$X(u, v) = \alpha(u) + \beta(v), \quad (u, v) \in I \times J,$$

with (I, α) and (J, β). This parametrization does not preserve the plane in which the second curve is contained as the plane at $\alpha(t)$ and normal vector $T_\alpha(t)$ for all $t \in I$.

Fig. C.28   QR Code 23. Exercise 4.15

Fig. C.29   Sweeping curve. Exercise 4.15

References

AAG. (2008). *Advances in Architectural Geometry*. Conference Proceedings of the First Symposium on Architectural Geometry.

AAG. (2010). *Advances in Architectural Geometry*, ed. C. Ceccato, L. Hasselgren, M. Pauly, H. Pottmann, and J. Wallner. Vienna, Austria: Springer.

AAG. (2013). *Advances in Architectural Geometry*, ed. L. Hesselgren, S. Sharma, J. Wallner, N. Baldassini, P. Bompas, and J. Raynaud. Wien/New York: Springer.

AAG. (2014). *Advances in Architectural Geometry*, ed. P. Block, J. Knippers, N. J. Mitra, and W. Wang. Springer Publishing Company, Incorporated.

AAG. (2016). *Advances in Architectural Geometry*, ed. S. Adriaenssens, F. Gramazio, M. Kohler, A. Menges, and M. Pauly. Vdf Hochschulverlag AG an der ETH Zürich.

AAG. (2018). *Advances in Architectural Geometry*, ed. L. Hesselgren, A. Kilian, S. Malek, K.-G. Olsson, O. Sorkine-Hornung, and C. Williams. Chalmers.

Anderson, S. (2004). *Eladio Dieste: Innovation in Structural Art*. Princeton Architectural Press.

Barrallo, J., and S. Sánchez-Beitia. (2011). The geometry of organic architecture: the works of eduardo torroja, felix candela and miguel fisac. *Proceedings of Bridges 2011: Mathematics, Art, Music, Architecture, Education, Culture*, 65–72.

Barrallo, J., A. González-Quintial, and Sánchez-Parandiet. (2018). Laminar reciprocal structures. *Proceedings of Bridges 2018: Mathematics, Art, Music, Architecture, Education, Culture*, 155–162.

Bärtschi, R., M. Knauss, T. Bonwetsch, F. Gramazio, and M. Kohler. (2010). *Wiggled brick bond*. Vienna: Springer.

Birindelli, I., and R. Cedrone. (2012). Modern geometry versus modern architecture. *Imagine Math*, 105–115.

Brander, D., A. Bærentzen, K. Clausen, Fisker, A.-S. Gravesen, M. N. Lund, T. B. Nørbjerg, K. Steenstrup, and A. Søndergaard. (2016). Designing for hot-blade cutting. In *Proc. AAG 2016*, 305–326.

Bridges. (2003). *Proceedings of Bridges Granada, Spain. Mathematics, Music, Art, Architecture, Culture*. Tessellations Publishing.

Bridges. (2004). *Proceedings of bridges Granada, Spain. Mathematics, Music, Art, Architecture, Culture*. Tessellations Publishing.

Bridges. (2008). *Proceedings of bridges Granada, Spain. Mathematics, Music, Art, Architecture, Culture*. Tessellations Publishing.

Bridges. (2011). *Proceedings of bridges Granada, Spain. Mathematics, Music, Art, Architecture, Culture*. Tessellations Publishing.

© The Author(s), under exclusive license to Springer Nature Switzerland AG 2021
A. Lastra, *Parametric Geometry of Curves and Surfaces*, Mathematics and the Built Environment 5, https://doi.org/10.1007/978-3-030-81317-8

Bridges. (2012). *Proceedings of bridges Granada, Spain. Mathematics, Music, Art, Architecture, Culture.* Tessellations Publishing.

Bridges. (2014). *Proceedings of bridges Granada, Spain. Mathematics, Music, Art, Architecture, Culture.* Tessellations Publishing.

Bridges. (2016). *Proceedings of bridges Granada, Spain. Mathematics, Music, Art, Architecture, Culture.* Tessellations Publishing.

Bridges. (2018). *Proceedings of bridges Granada, Spain. Mathematics, Music, Art, Architecture, Culture.* Tessellations Publishing.

Burden, R. L., and J. D. Faires. (2000). *Numerical Analysis.* Brooks/Cole.

Caliò, F., and E. Marchetti. (2015). Generation of Architectural Forms Through Linear Algebra. In *Architecture and Mathematics from Antiquity to the Future*, ed. K. Williams and M. Ostwald. Birkhäuser, Cham.

Capanna, A. (2012).   Architecture, form, expression. the helicoidal skyscrapers' geometry. *Proceedings of Bridges 2012: Mathematics, Music, Art, Architecture, Education, Culture*, 349–356.

Conversano, E., M. Francaviglia, M. Lorenzi, and L. Lalli. (2011).   Geometric forms that persist in art and architecture. *Proceedings of Bridges 2011: Mathematics, Music, Art, Architecture, Education, Culture*, 463–466.

Costa, A. F., J. M. Gamboa, and A. M. Porto. (1997). *Notas de Geometría diferencial de curvas y superficies*, ed. Sanz y Torres.

Čučaković, A., and M. Paunović. (2015). Cylindrical mirror anamorphosis and urban-architectural ambience. *Nexus Network Journal* 17:605–622.

de Burgos Román, J. (2006). *Álgebra lineal y geometría cartesiana.* McGraw-Hill Interamérica de España.

Dezeen (2021b). Nine examples of spherical architecture from around the globe, accessed 24 may 2021. https://www.dezeen.com/2020/08/29/spherical-architecture-buildings-roundup/.

do Carmo, M.-P. (1976). *Differential geometry of curves and surfaces.* Englewood Cliffs, NJ: Prentice-Hall, Inc.

Dolezal, J. (2011). The story of a right wavelet conoid. *WDS'11 Proceedings of Contributed Papers*, 72–77.

Dörrie, H. (2013). *100 great problems of elementary mathematics: their history and solution.* New York, NY: Dover.

Dunham, D. (2003).   Hyperbolic spirals and spiral patterns. *Proceedings of Bridges 2003: Mathematics, Music, Art, Architecture, Education, Culture*, 521–528.

Dzwierzynska, J., and A. Prokopska. (2018). Pre-rationalized parametric designing of roof shells formed by repetitive modules of catalan surfaces. *Symmetry* 10(4): 105.

Eastwood, M., and R. Penrose. (2000).   Drawing with complex numbers. *The Mathematical Intelligencer* No.4, 8–13.

Echevarría, L., C. Garnica, and J. P. Gutiérrez. (2014). La costilla laminar del Instituto de Ciencias de la Construcción Eduardo Torroja (IETcc-CSIC). Levantamiento mediante láser-escáner y evaluación estructural. *Informes de la Construcción* 66(536), e038. https://doi.org/10.3989/ic.14.116.

Effekt (2020a). Camp adventure. effekt, accessed 1 april 2020. https://www.effekt.dk/camp.

Emmer, M. (2013). Minimal surfaces in arquitecture: new forms. *Nexus Network Journal* 15:227–239.

Emmer, M. (2015). Architecture and Mathematics: Soap Bubbles and Soap Films. In *Architecture and Mathematics from Antiquity to the Future*, ed. K. Williams and M. Ostwald. Birkhäuser, Cham.

Erdös, P. (2000). Spiraling the earth with c. g. j. jacobi. *American Journal of Physics* 68:888–895.

Frazier, L., and D. Schattschneider. (2008).   Möbius bands of wood and alabaster. *Journal of Mathematics and the Arts* 2(3):107–122.

Gailiunas, P. (2014). Recursive rosettes. *Proceedings of Bridges 2014: Mathematics, Music, Art, Architecture, Education, Culture*, 127–134.

Gauss, C. F. (1876). *Werke, Zweiter, Band, Königlichen Gesellschaft der Wissenschaften.* Göttingen.

Gerbino, A. (2014). *Geometrical objects. Architecture and the mathematical sciences 1400-1800.* Springer International Publishing.

Gibson, C. (2001). *Elementary geometry of differentiable curves: An undergraduate introduction.* Cambridge: Cambridge University Press.

Glaeser, G., and F. Gruber. (2007). Developable surfaces in contemporary architecture. *Journal of Mathematics and the Arts 1(1)*, 59–71.

Glaeser, L. (1972). *The work of Frei Otto.* The Museum of Modern Art.

Gray, A. (1997). A different klein bottle. In *modern differential geometry of curves and surfaces with mathematica*, 327–330. Boca Raton, FL: CRC Press.

Hanh, A. J. (2012). *Mathematical excursions to the World's Great Buildings.* Princeton University Press.

Hart, G., and E. Heathfield. (2018). Catenary arch constructions. *Proceedings of Bridges 2018: Mathematics, Art, Music, Architecture, Education, Culture*, 325–332.

Hoffmann, C. M. (1989). *Geometric and solid modeling: An introduction.* San Mateo, CA: Morgan Kaufmann.

Inhabitat (2019c). Ooda's twisted cubic taipei city museum of art harvests rain and sun, accessed 20 may 2019. https://inhabitat.com/oodas-twisted-cubic-taipei-city-museum-of-art-harvests-rain-and-sun/.

Kilian, M., S. Flöry, Z. Chen, N. J. Mitra, A. Sheffer, and H. Pottmann. (2008). Developable surfaces with curved creases. *Proc. AAG*, 32–36.

Kimberling, C. (2004). The shape and history of the ellipse in washington, d.c. *Proceedings of Bridges 2004: Mathematics, Music, Art, Architecture, Education, Culture*, 1–12.

Klein, F. (2004). *Elementary mathematics from an advanced standpoint. Geometry.* Reprint of the 1949 translation. Dover Publications.

Koschitz, D., E. D. Demaine, and M. L. Demaine. (2008). Curved crease origami. *Proc. AAG*, 29–32.

Krivoshapko, S., and S. Shambina. (2012). Design of developable surfaces and the application of thin-walled developable structures. *Serbian Architectural Journal 4(3)*: 298–317.

Krivoshapko, S. N., and V. N. Ivanov. (2015). *Encyclopedia of analytical surfaces.* Cham: Springer.

Landsmann, G., J. Schicho, and F. Winkler. (2001). The parametrization of canal surfaces and the decomposition of polynomials into a sum of two squares. *Journal of Symbolic Computation* 32: 119–132.

Lang, S. (1986). *Introduction to linear algebra.* Undergraduate Texts in Mathematics. New York: Springer-Verlag.

Langbein, W., and L. B. Leopold. (1970). River Meanders and the Theory of Minimum Variance. In *Rivers and River Terraces. Geographical Readings*, ed. G.H. Dury. London: Palgrave Macmillan.

Lastra, A., and M. de Miguel. (2020). Geometry of curves and surfaces in contemporary chair design. *Nexus Network Journal*, 643–657.

Lastra, A., J. R. Sendra, and J. Sendra. (2018). Hybrid trigonometric varieties. Preprint. https://arxiv.org/pdf/1711.07728.pdf.

Lawrence, S. (2011). Developable surfaces: their history and application. *Nexus Network Journal* 13:701.

Liapi, K., A. Papantoniou, C. Noussias, and A. Ioannidi. (2019). Minimal surface tensegrity networks: the case of an enneper surface pavilion structure. In *IASS Annual Symposium 2019 –Structural Membranes 2019 - Form and Force.*

Mackin, N. (2016). Elliptic paraboloids in circumpolar vernacular architecture. *Proceedings of Bridges 2016: Mathematics, Music, Art, Architecture, Education, Culture*, 621–624.

Marchetti, E., and L. R. Costa. (2015). *What geometries in milan cathedral? architecture and mathematics from antiquity to the future*, vol. i, 509–534. Cham: Birkhäuser/Springer.

Marcus, J. W. (2008). Tensegrities. design, analysis and constructing. *Proceedings of Bridges 2008: Mathematics, Music, Art, Architecture, Education, Culture*, 389–392.

Marsden, J. E., and A. Tromba. (2012). *Vector calculus*. New York: W.H. Freeman.

Mathcurve (2019a). Courbe sphérique, accessed 20 may 2019. https://www.mathcurve.com/courbes3d/spheric/spheric.shtml.

Miltra, N. J., and M. Pauly. (2008). Symmetry for architectural design. In *Proc. AAG*, 45–48.

Motro, R. (2003). *Tensegrity: structural systems for the future*. London; Sterling, VA: Kogan Page Science.

Munkres, J. R. (1974). *Topology; a first course*. Englewood Cliffs, NJ: Prentice-Hall.

Nadenik, Z. (2005). Lame surfaces as a generalisation of the triaxial ellipsoid. *Studia Geophysica et Geodaetica* 49:277.

Narváez-Rodríguez, R., and J. A. Barrera-Vera. (2016). Lightweight conical components for rotational parabolic domes. In *Proc. AAG*, 378–397.

Narváez-Rodríguez, R., A. Martín-Pastor, and M. Aguilar-Alejandre. (2014). The Caterpillar Gallery: Quadratic Surface Theorems, Parametric Design and Digital Fabrication. In *Advances in Architectural Geometry*, ed. P. Block, J. Knippers, N. Mitra, and W. Wang. Cham: Springer.

Nikolić, D., and V. Živaljević. (2020). On the modelling of vaulted structures of equal strength. *Nexus Network Journal*, 1219–1236.

Peternell, M., and H. Pottmann. (1997). Computing rational parametrizations of canal surfaces. *Journal of Symbolic Computation* 23:255–266.

Picon, A. (2010). *Culture numérique et architecture : une introduction*. Editions Birkhäuser.

Postle, B. (2012). Methods for creating curved shell structures from sheet materials. *Buildings* 2:424–455.

Pottmann, H., A. Asperl, M. Hofer, and A. Kilian. (2007). *Architectural geometry*. Bentley Institute Press.

Pottmann, H., and J. Wallner. (2001). *Computational line geometry*. Springer Verlag.

Rossi, A., and U. Palmieri. (2020). Modelling based on a certified level of accuracy: The case of the solimene façade. *Nexus Network Journal*, 615–630.

Sagan, H. (1992). *Introduction to the calculus of variations*. Courier Dover Publications.

Sala, N. (2003). Fractal geometry and self-similarity in architecture: an overview across the centuries. *Proceedings of Bridges (2003): Mathematics, Music, Art, Architecture, Education, Culture*, 1–10.

Salas, S. L., E. Hille, and G. J. Etgen. (2003). *Introduction to the calculus of variations*. Calculus: One and several variables. New York: J. Wiley & Sons.

Samyn, P. (2017). Between light and shade, transparency and reflection. https://samynandpartners.com/17_e-books/ombreetlumiere-EN/mobile/index.html.

Samynandpartners (2020b). Samyn and partners. architects & engineers, accessed 1 april 2020. https://samynandpartners.com/fr/portfolio/walloon-branch-of-reproduction-forestry-material/.

Schiftner, A., J. Raynaud, N. Baldassini, and H. Pottmann. (2008). Architectural freeform structures from single curved panels. In *Proc. AAG*, 45–48.

Schling, E., M. Kilian, H. Wang, J. Schikore, and H. Pottmann. (2018). Design and construction of curved support structures with repetitive parameters. In *Proc. AAG*, 140–165.

Sendra, J., F. Winkler, and S. Perez-Diaz. (2007). *Rational algebraic curves: A computer algebra approach. Series: algorithms and computation in mathematics*, Vol. 22. Springer Verlag.

Séquin, C. H. (2008). Algorithmically acquired architectural and artistic artifacts. In *Proc. AAG*, 9–12.

Séquin, C. H. (2018). Möbius bridges. Journal of Mathematics and the Arts. Académie Royale de Belgique. Series Pocket Book Academy.

Sharpe, T. (2010). *The role of aesthetics, visual and physical integration in building mounted wind turbines - An alternative approach, paths to sustainable energy*. Jatin Nathwani and Artie Ng. IntechOpen.

Shen, Y., E. Zhang, Y. Feng, S. Liu, and J. Wang. (2021). Parameterizing the curvilinear roofs of traditional chinese architecture. *Nexus Network Journal*, 475–492.

Sperling, G. (2015). The "quadrivium" in the pantheon of rome. *Architecture and mathematics from antiquity to the future*, vol. I, 509–534.

Strang, G. (1993). *Introduction to linear algebra.* Wellesley-Cambridge Press.

Tang, C., M. Kilian, P. Bo, J. Wallner, and H. Pottmann. (2016). Analysis and design of curved support structures. In *Proc. AAG*, 8–23.

Tapp, K. (2016). *Differential geometry of curves and surfaces.* Undergraduate Texts in Mathematics. Springer.

Tellier, X., L. Hauswrith, C. Douthe, and O. Baverel. (2018). Discrete cmc surfaces for doubly-curved building envelopes. In *Proc. AAG*, 165–193.

Thulaseedas, J., and R. J. Krawczyk. (2003). Möbius concepts in architecture. *Proceedings of Bridges (2003): Mathematics, Music, Art, Architecture, Education, Culture*, 353–360.

Tofil, J. (2007). *Application of Catalan surface in designing roof structures – an important issue in the education of a future architect engineer*, Vol. 14(137). International conference on Engineering Education-ICEE 2007, September 3–7, Coimbra, Portugal.

Torroja, E. (1962a). Hipódromo de la zarzuela. *Informes de la Construcción*, Vol. 14(137).

Torroja, E. (1962b). La cuba hiperbólica de fedala. *Informes de la Construcción*, Vol. 14(137).

Umehara, M., Y. Kotaro, and W. Rossman. (2017). *Differential geometry of curves and surfaces.* World Scientific.

Velimirovic, L., G. Radivojevic, M. Stankovic, and D. Kostic. (2008). Minimal surfaces for architectural constructions. *Facta Universitatis - Series: Architecture and Civil Engineering*, 6(1):89–96.

Vergopoulos, S. (2010). Design intentions and innovation new teaching paradigms in the context of digital architectural design. educating architects towards innovative architecture. *Proceedings of the 46th ENHSA-EAAE Architectural Design Teachers' Network Meeting, Istanbul - Turkey*, 411–422.

Wallner, J., A. Schiftner, M. Kilian, S. Flöry, M. Höbinger, B. Deng, Q. Huang, and H. Pottmann. (2010). Tiling Freeform Shapes With Straight Panels: Algorithmic Methods. In *Advances in Architectural Geometry*, ed. C. Ceccato, L. Hesselgren, M. Pauly, H. Pottmann, and J. Wallner. Springer, Vienna.

Wikiarquitectura (2021a). Klein bottle house, accessed 10 february 2021. https://en.wikiarquitectura.com/building/klein-bottle-house/.

Wikipedia (2019b). List of hyperbolid structures, wikipedia.com, accessed 30 april 2019. https://en.wikipedia.org/wiki/List_of_hyperboloid_structures.

Index

© The Author(s), under exclusive license to Springer Nature Switzerland AG 2021
A. Lastra, *Parametric Geometry of Curves and Surfaces*, Mathematics and the Built
Environment 5, https://doi.org/10.1007/978-3-030-81317-8

Printed in the United States
by Baker & Taylor Publisher Services